BEE TIME

BEE TIME

LESSONS FROM THE HIVE

Mark L. Winston

Harvard University Press

Cambridge, Massachusetts
London, England

Printed in the United States of America

First Harvard University Press paperback edition, 2016
Second printing

Library of Congress Cataloging-in-Publication Data
Winston, Mark L.
  Bee time : lessons from the hive / Mark L. Winston.
    pages cm
  Includes bibliographical references and index.
  ISBN 978-0-674-36839-2 (cloth : alk. paper)
  ISBN 978-0-674-97085-4 (pbk.)
  1. Bee culture—Social aspects. 2. Honeybee—Social aspects.
3. Honeybee—Behavior. 4. Bees (Cooperative gatherings) I. Title.
  SF523.3.W547 2014
  638'.1—dc23   2014008698

For Lori Bamber

Contents

BEE TIME

Prologue

Walking into the Apiary

Walking into an apiary is intellectually challenging and emotionally rich, sensual and riveting.

Time slows down. Focus increases, awareness heightens, all senses captivated.

Entering an apiary has its own rhythm and ritual. I slip my pants into my boots, put on my veil, light the smoker to calm the bees, all routine preparations imbued with deeper meaning because they herald the transition from whatever I had been doing into bee mode.

Lifting my smoker, I am totally in the present but also connected to memories of friends, fellow beekeepers, and innumerable long days in other apiaries when we shared periods of tedium, hard physical labor, and occasional glimpses of wisdom. These moments of understanding, penetrating the complexity of our usually unfathomable natural world, still take my breath away.

I remember my first breathtaking moment of revelation, a much-anticipated apiary visit that turned out to be nothing like I had expected. I had begun graduate school at the University of Kansas, choosing the Department of Entomology because it had a history of sending students to the tropics for their research, and I was determined to become a tropical biologist. My advisor had just received a grant to study African honeybees in South America. I heard "South America," thought bees would do just fine, and soon found myself in French Guiana in July 1976 entering an apiary of killer bees.

I had completed an exhausting trip only the day before, from Kansas to Miami to Martinique to Cayenne, French Guiana, and was disoriented. My experience with bees up to then consisted of a few brief trips to the university's local apiaries, and I had no idea what to expect. I entered this South American apiary wearing a layer of clothes under a bee suit, my head covered by a veil, my hands inside gloves, believing I was fully protected from what I thought would be an assault of wrathful bees.

I removed the lid of the first hive. The bees were surprisingly calm, gentle, going about their business. There was no onslaught. My fear dissipated, and I began to pay attention to the activity in the hive. The gloves came off, then the veil. I pulled out the frames one by one to inspect the combs.

It was one of those moments in life when everything shifts.

It's a full-body experience being among the bees. First you hear the sound, the low hum of tens of thousands of female workers flying in and out of their hives, each circling the apiary to get her bearings and then heading off purposefully in a literal beeline toward blooming flowers. Smells and textures bombard the senses next, the sweet odors of beeswax and honey, the stickiness of plant resins collected by the workers to plug holes and construct the base of their combs. And then there are the bees themselves, walking over your hands and forearms as you lift and return combs from the hive, the subtlest of touches as their claws lightly cling and release, the gentlest of breezes as their wings buzz before taking flight.

A first look at the hive reveals it to be a place of activity and complexity. Some bees have their heads in the comb's cells feeding larvae. A few are fanning their wings to evaporate water in the honey. Others are performing the waggle dance, alerting their nest mates to the location of nectar and pollen in the landscape. Many are immobile, perhaps resting or waiting for a new task to present itself.

The most common activity is interaction: two bees frenetically touching antennae, legs, and tongues, climbing over each other and doing the same with the next bee they encounter. Occasionally you glimpse the queen, stately and slowly moving across the comb, surrounded by workers in attendance, inserting her abdomen into a cell about once a minute to lay an egg. Underlying all the physical sensations are collaboration and order, communication and common purpose, each bee submerging her individual nature for the colony.

I have spent a considerable portion of my life in apiaries in Canada and the United States, New Zealand and Australia, South America and Europe. Some have been deep in jungles surrounded by tropical vegetation and the calls of birds and monkeys. Others have been in northern alpine meadows where colonies had been moved for a brief few weeks to forage on purple-spiked fireweed flowers, which yield the clearest of water-white honey.

Many of these apiaries have been adjacent to stunningly yellow canola fields blooming on the Canadian prairies, where bees forage at midnight under the summer solstice sun. Still others have been tucked behind fences and near gardens in urban backyards, small oases of escape from the cacophony of urban life outside the apiary.

No matter whether in a jungle or city, next to a freeway or by the most scenic of creeks, fields, forests, or weedy vacant lots, entering an apiary has never failed to engage my senses and focus my attention.

In these places I learned powerful lessons from the bees about how we humans can better understand our place in nature, engage people and events with greater focus and clarity, interact more intensely in our relationships and communities, and open our hearts and minds to a deeper understanding of who we are as individuals, communities, and a species.

Not then, not until long after I had left the apiaries behind did I come to think of these apiary moments as "bee time." I had become the director of my university's Centre for Dialogue; my work had evolved into realms far from bees. I was facilitating discussions of the complex and nuanced issues that face contemporary human societies. The settings were classes and workshops, large public dialogues, and private one-on-one conversations, sometimes focused on adversarial and controversial public issues, sometimes on the most deeply personal and intimate reflections.

A few years after I had moved into this new world of human interface I was interviewed by a journalist who noted that bees and dialogue didn't seem connected and wondered whether they had anything in common. Absolutely, I responded. Initiating a dialogue requires the same attention as entering an apiary. Both stimulate a state of deep listening, engage all the senses: hearing without judging.

Through dialogue, time slows down, as it does in apiaries. Focus sharpens on how participants are interacting. Understandings emerge, issues clarify and become connected, and collaboration surfaces from the intentions and actions of many individuals. Solitary becomes communal.

Dialogue has that apiary feeling, reading situations and discerning what there is to learn from each unique constellation of persons, circumstances, and issues.

Those too-rare moments of presence and awareness, when deep human interactions are realized: they, too, are bee time.

1

Beginning with Bees

Bees have been entwined with our history since the appearance of the earliest humans, but bees were here long before us. The first bees evolved from wasps about 125 million years ago, shifting from predators to gatherers of nectar and pollen from flowers.

These early bees were solitary, nesting in hollow twigs or in soil, and were notable for branched hairs that trapped pollen grains when a bee visited a flower. From these hairs the pollen was (and still is) transferred to other flowers on subsequent visits, thereby pollinating (fertilizing) the plant's seed.

Flowers secrete nectar and/or excess pollen as a food reward to interest the bees, the nectar providing carbohydrates and the pollen protein. This evolutionary innovation led to an explosion in the diversity and abundance of advanced plants, coinciding with that of bees, and eventually to the biosphere we know today.

More than twenty thousand known species of bees currently live on every continent except Antarctica, displaying a breathtaking array of lifestyles. Some are solitary; others live in loose communal groups or form highly complex societies. They nest in sandy soil, hard dirt, abandoned rodent nests, and hollow twigs, stems, and tree trunks. Their flower-visiting habits range from omnivorous, taking nectar and pollen from a wide range of flowers, to obligate relationships, in which a plant species can be pollinated by only one bee species, which in turn visits only that plant's flowers.

These diverse species are vitally important to human economies and to the world's welfare because of their pollinating role in both agricultural and natural ecosystems. Their global economic value for agriculture was estimated at US$217 billion in 2008, with about one-third of all crops benefiting from or dependent on insect pollination, mostly by bees. Without bees we would have a vastly diminished grocery, missing most of the fruits, vegetables, berries, and nuts that we depend on for a healthy, balanced diet.

Bees' value to natural ecosystems as pollinators is incalculable. We can price the pollination services that nature provides for crop production, but it's more challenging to monetize the full range of services provided by the rich palette of plants that depend on bees. The dependency of other organisms on bee-pollinated plants is broad, including food and shelter for innumerable feral animals, the conversion of carbon dioxide into oxygen, which maintains the earth's life-supporting atmosphere, the role of plants in stabilizing soil and preventing erosion, and many other functions.

Perhaps one way to assess the worth of bees for nature is to consider the extent to which advanced plants (angiosperms) depend on them for pollination. At the University of Northampton, Jeff Ollerton, who studies ecosystem valuation, reports that 20 percent of all angiosperms can be pollinated only by bees, while another 45 percent are pollinated by bees as well as by the wind or other animals, mostly beetles, moths, butterflies, birds, and bats. That is, 65 percent, or 229,000 of the 352,000 species currently known, require or benefit from bee pollination.

Clearly, any significant drop in bee populations will have repercussions that go far beyond the loss of bees. A world without bees would be almost impossible to contemplate and likely one in which we would never have evolved in the first place.

The array of wild bees is staggering, with lifestyles and habits that fascinate any lover of natural history. Among my personal favorites are the giant orchid bees, huge metallic bees that make a deafening buzzing noise as they wend their way through tropical jungles seeking the specific orchid flowers that each species pollinates. One of their most interesting habits is scent collection from flowers by the males, which then use these scents to attract females.

At the other end of the earth's climate zones are the bumblebees, common in temperate latitudes, found high up in alpine zones and within six hundred miles of the North Pole, at the northernmost points of land adjoining the Arctic Ocean. These bees have superbly adapted to cold climates, flying in freezing temperatures and even snow by disconnecting their wings and heating up their flight muscles to high temperatures by shivering. They nest in colonies with a queen and a few hundred to a few thousand workers. Each colony dies every fall; new queens, which are produced in late summer, leave the nest, find protected cavities in which to overwinter as solitary individuals, and then found their own colonies in the spring.

The stingless bees are another unusual group, five hundred or so tropical and subtropical species that, over evolutionary time, have lost their stings. These social bees build large nests in hollow tree trunks or underground cavities, where they store nectar and pollen and rear their young in egg-shaped pots constructed from wax they produce in special glands. Although they cannot sting, they are hardly defenseless, having evolved jaws and associated venom sacs that deliver bites as least as painful as the stings of other bee species.

Less heralded but ecologically crucial are myriad solitary bees, generally small and unobserved but providing the blue-collar pollination work required for ecosystems to thrive. Only the keenest observers will see these minute bees darting from flower to flower, returning to provision a few eggs with a paste of nectar and pollen before sealing the nests to let their young mature, hidden from predators and parasites.

But when we think of "bee," it's usually honeybees that come to mind since they are certainly the bees with which we have had the most interaction. They are gloriously social, producers of honey, source of much art and story, and deeply ingrained in many aspects of human economy and lore. We are intimately connected; our and their prosperity is closely linked to each other's well being.

o   o   o

What is unique about our relationship with honeybees is not only how much we depend on their services but also the fact that their health and survival depend on how well we manage the environment on which they rely. If we had a formal contract with honeybees, its executive summary might read something like this: We, the bees, will provide you with honey and other products of the hive, as well as pollination services. In return you, the humans, will maintain an environment in which we can thrive, free of toxic pesticides and rich in diverse flowering plants.

Sadly, our implicit arrangement with honeybees is becoming increasingly frayed. Their health has been challenged by the toxins we release into the environment, agricultural practices that have severely diminished the diversity and abundance of the nectar and pollen sources upon which they depend, our transport of bee pests and diseases around the globe, and beekeeping practices that overly stress managed honeybee colonies. And we, in turn, are threatened with diminished crop yields and reduced/tainted honey supplies as a direct result of our disregard for our honeybee partners.

The current dangers facing this key species are particularly tragic given their long evolutionary history and our long-standing interactions. The first honeybees appeared in the fossil record about forty million years ago, although they likely evolved much earlier, and by thirty million years ago were essentially the same as the honeybees we see today.

Currently ten to twelve honeybee species exist, mostly in Asia, but the best known is the Western honeybee, *Apis mellifera*, with its original geographic range from Africa to Europe but now found globally due to human transport of colonies. Its scientific name means "honey-bearing" or "honey-producing" bee, referring to its singular capacity to collect copious quantities of nectar and store it as honey so that colonies can survive dearth periods.

Storing honey was a significant adaptation for honeybees, a trait that evolved simultaneously with perennial nests constructed from beeswax in which to house the honey, as well as the extraordinarily sophisticated social behaviors equally important to honeybee success. But it's the honey that drew us to them initially, as well as the larvae, which are replete with protein, both food sources for early humans who would raid wild nests and suffer stings in order to extract these rich culinary rewards.

Even then I imagine that we were intrigued with honeybees, perhaps sensing a peer among the animals, one of the few with which we share a deeply ingrained sociality. Still, it was the advent of managed hives that provided an opportunity for deeper glimpses into honeybee society and surely increased our curiosity and fascination.

The first hives were nothing more than hollow logs hung from trees to catch swarms, with open ends through which early beekeepers could reach in and remove honey-laden combs while also observing the bees at work. As hives became more artificial, constructed from mud, clay, or woven baskets, our opportunities to observe bees increased. But it was the invention of movable frames in 1860 by Philadelphia clergyman Lorenzo Langstroth that resulted in a significant increase in colony productivity and augmented our capacity to observe the complex social mechanisms that hold honeybee colonies together.

Langstroth's movable frames hang in the rectangular boxes that make up today's familiar beehives, spaced at just the right intervals to allow the bees to build comb in each frame while providing sufficient room between the frames to move and work. Each one can be pulled out, inspected, and returned to the hive. Honey is removed by spinning heavily laden frames in a centrifuge-like extractor; the honey is flung to the inner extractor walls and flows down for bottling, and the now-empty combs are then placed back in the hive.

The Langstroth hive dramatically increased honey yields for a number of reasons. First, beekeepers could now stack boxes, thereby organizing colonies with the nursery below and storage areas above, increasing the area for the bees to store honey while also keeping it separate from the brood so that it was easier to extract. Also, the frames could be reused after extraction, a significant advantage because it takes considerable energy for the bees to construct their comb. In addition, movable frames allowed beekeepers to inspect colonies for disease and the presence of a queen and take remedial action when necessary.

The advent of movable frames had another consequence; scientists could now observe and experimentally manipulate colonies for study, which led to a cascade of discoveries that continue today. Glass-walled observation hives further opened the world of honeybees to surveillance, while colored and numbered labels glued to the backs of bees permitted us to trace the behaviors of individual workers throughout their lifetimes.

Our observations and analyses of honeybees over the last many millennia accelerated during the past 150 years with the modern hives, resulting in a remarkable increase in knowledge and information. Essentially, modern science has only confirmed what our ancestors knew intuitively: honeybees are as sophisticated in their social behavior as we are in ours, sharing with us a bond that is uncommon among the earth's many creatures.

o   o   o

Honeybees fascinated early scientists and naturalists, although their keen interest did not always yield correct observations. Aristotle, for example, believed the female queen to be a male king and was convinced that the worker bees developed from some progenitor substance that adult workers brought back from flowers rather than from eggs laid by the monarch. He also believed that honeybee comb was constructed from tree resins, missing the concept of the bee-produced wax (beeswax), which we now know to make up the comb.

Still, Aristotle got a fair bit right. He noted the four life stages of honeybees—eggs, larvae, pupae, and adults—and surmised that worker bees feed the larvae. He also understood the remarkable insect transformation that changes a larva into an adult through a pupal stage, and he accurately deduced that this metamorphosis takes place in a cell sealed by adult workers over a twelve- to thirteen-day period.

Perhaps the most astute natural observer of honeybees was the Swiss naturalist François Huber, who at the turn of the eighteenth century correctly elucidated many aspects of colony life history. Huber's observations are particularly compelling because he went blind at the age of fifteen, relying on his wife and a servant to observe bees and supply him with information that his astute scientific mind then translated into hypotheses about honeybee behavior.

Huber understood that the monarch in the hive was a female, not a male, as Aristotle believed. He recognized that honeybees reproduce by swarming, when the old queen and a majority of the workers leave the nest to seek a new site.

Huber also observed that the developing queen larvae are left behind and soon emerge as adults to fight other unmated queens to the death until only one is left. Then, the last virgin standing leaves the hive to mate, copulating with ten to twenty male drone bees that catch the queen and then die following their aerial tryst. The mated queen returns to the nest and begins laying eggs within a few days, reestablishing order in the colony.

Today we understand considerably more than Aristotle and Huber, and our deepening knowledge of honeybees has only increased our fascination. We know that colonies generally consist of only one queen that influences but doesn't command colony activities. In fact, researchers and keenly observant beekeepers have learned that the term "queen" isn't quite as royal for bees as for human monarchs since governance in honeybee colonies is highly decentralized. Worker bees have considerable decision-making capacity based on their assessment of the localized colony conditions they encounter and information about the outside world that foragers bring back to the hive. The idea of a powerful reigning monarch familiar to traditional human societies doesn't translate well into the honeybee world.

The queen does have one major task, laying eggs, and all the worker bees in the colony are her daughters. Physically the workers are quite distinct from the queen: they are smaller and manifest specialized structures useful in accomplishing their many tasks. These include expanded legs with stiff hairs to collect pollen and a concave basket on the hind legs in which to carry it back to the nest, glands in the abdomen that secrete wax the workers use to build comb, and a well-developed barbed stinger for defense.

The hive is rounded out by hundreds of the queen's sons, called drones, whose sole function is to mate with virgin queens from nearby colonies. The drones do no work and are driven out of the colony to die every fall when the mating season ends. The next spring they are replaced by new drones.

Each egg laid by the queen has the potential to develop into any one of the three castes. The egg will develop as a female if fertilized with sperm held in a sac near the queen's ovaries. If fed extra portions and a particularly nutritious diet, the young larva that hatches from the egg will mature into a queen; a regular diet results in a worker bee. If the queen's egg is not fertilized, then a drone results. Occasionally a worker bee, which has small nonfunctional ovaries, will expand her ovaries and lay an egg that develops into a drone, although this is unusual unless the queen has died and the colony has failed to rear a replacement queen.

As fascinating as individual bees may be, it is the colony that intrigues us the most. Honeybees, like us, are challenged to balance the role of individuals relative to the communal good. On a continuum between the personal and the collective, bees very much favor their community. Although we may or may not decide to emulate their social conscience, comparing ourselves to honeybees serves as a useful reflector for contemplating our own social nature.

o o o

Our interest in honeybees rests on the twin pillars of their sociality and their economic value, which makes their current plight particularly poignant. Not only is a mainstay of agricultural productivity in decline, but honeybees are also the most highly social of all our domesticated animals. As we watch their societies collapse, we can envision our own demise.

The loss of honeybees and a parallel decline in wild bee populations reflect challenges similar to those we face: an increasingly complex set of environmental perturbations that eventually reach a tipping point, beyond which survival is problematic.

For honeybees, this condition has been called *colony collapse disorder,* a syndrome in which virtually all of the bees disappear within days or weeks. A beekeeper visits one week and all looks well, but by the next visit a few weeks later the colony is dead. This is a global phenomenon that afflicts about one-third of honeybee colonies each year, a catastrophic rate for both bees and their managers.

Wild bees have been less studied, but there is considerable evidence that their populations are declining as well, and some once-common species are now rare or have even become extinct. Certainly their populations in and around cropland are sparse and insufficient for pollination. As a result, growers of bee-pollinated crops import hired honeybees during bloom to provide a managed service that ecosystems used to perform naturally and free of charge. The decline of honeybees has left farmers scrambling, and we're on the verge of serious declines in crop yields if we do not address pollination issues quickly.

Bee decline is not caused by a single factor, and it is this example of complex causation that should be of greatest concern to us. For honeybees, the precipitous drop in colony numbers in the last eight to ten years is attributed to a perfect storm of many factors. Pest and disease outbreaks are rampant, with mites, bacteria, viruses, and fungi decimating colonies. Their impact on honeybees is exacerbated by diminished immune responses caused by pesticides, both those spread externally in farmers' fields and those applied inside colonies as beekeepers attempt to chemically control the outbreaks. Chemical overuse by beekeepers themselves has resulted in resistant pests and diseases, compounding the problem.

Reduced diversity and abundance of nectar- and pollen-producing flowers afflict both honeybees and wild bees, a direct impact of contemporary mass agriculture. Farmers' fields have become wastelands for bees: vast acreages of a single crop bloom for a week or two and then turn into a floral desert due to herbicide treatments that clean the fields of the blooming weeds, which used to provide the mixed nectar and pollen sources essential for adequate bee nutrition.

And widespread insecticide use kills bees as well as pest insects, resulting in direct damage particularly to wild bees, which, unlike honeybees, can't be moved away from fields during spraying. In addition, plowing and planting disrupt potential nesting sites for wild bees, further challenging these vulnerable creatures.

It's a grim picture and a particularly compelling one as we face a blend of similarly far-reaching and complex global perturbations that threaten our human populations. The growing effects of climate change, health challenges from exposure to a vast array of pesticides and industrial pollution, diminished access to diverse foods in many parts of the globe, and increased outbreaks of bacterial diseases resistant to antibiotics all seem eerily reminiscent of the plight bees find themselves in.

Whether we're about to collapse, as honeybees have, is arguable, but learning from their plight provides insights into how we might best face our human challenges. What's most remarkable is not that bees are dying but that they have survived and thrived until recently in increasingly inhospitable environments.

○ ○ ○

But this is not a book filled only with gloom; the resilience of bees is more inspirational than depressing and provides considerable promise that we, too, can learn to work together and thrive. There is much to celebrate about our long relationship with bees and many positive and uplifting insights that impart lessons more filled with hope than doom.

In the chapters that follow we will explore the calamity of bee demise but also discover some habitats in which bees are unexpectedly thriving, particularly cities. We'll see that there are practical, economically viable steps in how we manage wild bees and honeybees and agricultural ecosystems, with excellent potential to create an environment more suitable for bees and us as well.

We'll discuss honey, that marvelous product of the hive, learn how bees produce it, and discover some of its splendid properties. We'll see the best and the worst of the honey business, including encounters with the finest of artisanal beekeepers and the largest case of global food fraud ever uncovered.

Bees have a rich history of inspiring spiritual practices and art in virtually all forms of media from painting to sculpture, poetry to film, music to dance. We'll visit shamanic beekeepers, ponder the mindset behind deliberately getting stung to cure diseases, learn how bees inspired the earliest painters who drew on cave walls, and hear from a renowned artist who inserts everyday objects into hives to transcend the communication barrier between us and the bees.

We'll also delve into the latest scientific breakthroughs probing the mechanisms underlying honeybees' essential sociality. Although we may have stressed honeybees to their breaking point, they are spirited and irrepressible, buffered from the world around them by their collaborative nature and the community-first decisions made by individual bees.

If there is one notable message from honeybees, it lies in the power of their collective response to stress, in the way they allocate work, communicate, make decisions, and balance individual activities with their communal imperatives. Our decision either to emulate honeybees by opting for the collective good or to pursue personal interests and individual gain may be the decisive factor in the success or failure of our response to contemporary environmental challenges.

We have much to learn from bees. Let's begin by opening an unusual bottle of mead made from the honeybees' most valued product, honey.

2

Honey

One of the occupational hazards of being a bee expert is that beekeepers often ask me to sample their honey and homemade mead. I have had a lot of pretty good honey but haven't been so fortunate in the mead department.

Mead can be pretty tough to quaff, whether store bought or homegrown. It simply is not that easy to produce a superior honey wine. Those of us who are invariably polite tremble when the bottle of home brew is brought up from the basement by our hosts, terrified that we are about to enter the undrinkable zone. The rarity of a good bottle of mead makes those experiences memorable.

One remarkable example was sampled at my own table, poured from a bottle that had been in my basement for more than ten years—and looked it. Covered with dust, a bit of mold growing on the top of the cork, a layer of turbid sediment at the bottom, it was labeled with a peeling piece of masking tape proclaiming that this was "Atkinson Mead by HARVEY BOONE 1968."

I had feared uncorking this bottle since the moment it was given to me as a gift because the giver was John Boone, one of the nicest of gentlemen in a community not generally known for its gentleness—beekeepers. The mead maker was his father, Harvey, an icon in British Columbia beekeeping.

John, a cardiologist, had given me this relic of Canadian beekeeping history after attending a course at the university where I was teaching. Harvey Boone was a founding father of beekeeping in British Columbia. Even today, a few generations after his time, he is still revered by our beekeepers; a trust fund that sponsors research and education was named in his honor. Ted Atkinson, a scientist in the Okanagan Valley fruit-growing region, was recognized on the label because he provided Harvey with some apple concentrate to juice up the recipe.

I kept putting off opening Boone's mead. A precious bit of beekeeping history lay in my basement and I was deeply apprehensive that it would taste pretty much like it looked. Family celebrations came and went, as did innumerable guests from the beekeeping world who graced our table and would have appreciated the history behind the heirloom.

Years passed, the bottle got dustier and dustier, but finally the perfect occasion arrived. It was August 2003, and our close apicultural friends Don Dixon and his wife, Jamie, were staying with us. My daughter who was home from university joined us at dinner with her best friend from high school.

Rounding out the table was Michael Young from Northern Ireland, a cook of considerable repute who oversees a culinary institute at which many of Ireland's finest chefs train. He also happens to be one of the world's premier mead makers. I had spent two nights at his home the month before, prior to speaking at an Irish Federation of Beekeepers event, in a large spare room filled with the trophies he has garnered for his mead.

Harvey Boone's mead called from the basement. I dusted off the bottle as best I could, passed it over to Michael to open, and waited for my worst fears to be realized. The cork popped quietly, the tawny mead was poured into our best crystal aperitif glasses. The moment had come.

It was sublime. Beyond sublime, it tasted like nothing I had ever experienced. Deep, rich, smooth, with a hint of the fruity honey and concentrate from which it was fermented, this was a brilliant bottle of mead. Humble its container may have looked, but within was a flavorful jewel, an experience worthy of the producer's exemplary reputation.

Boone's mead is gone, but I kept the bottle as a reminder of a special night when good mates raised glasses to toast friendship and the achievements of a great beekeeper. It still holds the scent of Boone mead, an aromatic memory evoking a moment when beekeeping past and present gathered in our home.

o   o   o

Beekeeping and honey do that, attracting friends and family around the table to reflect on the mystique embodied in food given to us by honeybees, a species that existed on Earth almost forty million years before humankind arrived. And there still is much artisan honey to be found and enjoyed, imbued with the charisma of this marvelous bee offering.

But honey also has its dark side, its purity challenged by industrial processing, underground trading in unlawful honey, mislabeling of floral source and country of origin, adulteration with corn syrup, and contamination with banned pesticides and antibiotics. That pristine-looking jar of honey on the supermarket shelf may have a criminal history, and the stories of honey reveal both the best and the worst elements of contemporary food culture.

For bees, though, honey is about survival rather than gastronomy. Bees obtain virtually all of their energy from plant-produced nectar, a complex, watery soup of sugars. Other substances in nectar are important for bee nutrition, particularly minerals, vitamins, lipids, and ascorbic acid, but they are present in only minute quantities. Nectar also contains aromatic floral odors that attract the bees and provide the unique flavor and aroma of each plant's nectar.

The deal is pretty basic, although the details can be elegant. Flowers provide sugar in nectar as well as protein in pollen, meeting the two principal food requirements for bees. For their part, bees transfer pollen from flower to flower as they collect their rich rewards, thereby fertilizing the flower.

Kahlil Gibran, in his 1923 book, *The Prophet,* commented on the more spiritual dimensions of this relationship between bee, flowers, and honey: "Go to your fields and gardens, and you shall learn it is the pleasure of the bee to gather honey of the flower. But it is also the pleasure of the flower to yield its honey to the bee. For to the bee a flower is the fountain of life. And to the flower a bee is a messenger of love."

Honeybees are unusual among bees in processing large quantities of nectar into the storage form of honey. Floral nectar is carried back to the nest in the worker bees' specialized second stomach, where enzymes are added that break down the complex forms of sugar in nectar into simpler, inverted forms that are easier for bees, and us, to digest. These simple forms of sugar are also better at resisting bacterial attack than more intricate sugars. Other enzymes catalyze the production of tiny amounts of at least two and probably more distinct compounds, hydrogen peroxide and methylglyoxal (MGO), which together prevent bacterial and fungal degradation.

The foraging honeybees regurgitate the now-processed nectar when they return to the hive, evaporating some of the water by layering it thinly on their tongues and then placing it into cells. Younger house bees stand over the nectar and fan it with their wings, reducing the water content of the nectar to 18.6 percent or less to protect the honey from yeasts. At this point the nectar is "ripened" and can be called honey.

Finally, comb-building bees secrete wax and put a capping over the cell for long-term storage. And long-term it is; honey—still edible—has been found in the tombs of pharaohs.

A typical colony in temperate climates requires 130–180 pounds of honey annually to survive. To collect each pound, tens of thousands of bees fly about fifty-five thousand miles in total and visit more than two million flowers.

The essence of beekeeping is that beekeepers remove much of the stored honey every summer and fall, replacing it with sugar syrup they feed to the bees before winter. And they remove a lot of it: According to the Food and Agriculture Organization, beekeepers harvested 3.4 billion pounds from thirty million colonies worldwide in 2010, an average global yield of 113 pounds per colony.

What we do with that honey speaks volumes about how we view food and the bounty of sustenance that nature provides for us.

o o o

Honey at its best is of the land, a signature of time and place reflected through the distinctive odors and flavors of the terrain through which bees forage.

It's the essence of *terroir,* a French term originally used to indicate how the taste of a bottle of wine is affected by ecological factors in local vineyards. Wine connoisseurs discern the effects of local soil conditions, weather, and grape variety in each bottle's bouquet. Deep *terroir* aficionados insist that wine made from grape rows high up on a hill taste distinctively different from wine made with grapes even a few rows lower in the vineyard due to microclimatic differences in sunlight, heat, wind, and soil composition.

Honey also reflects the environment to us but represents a wider sampling of the landscape than a fixed-location crop like grapes. An apiary of twenty to thirty colonies takes up only a few dozen square yards, but its foraging bees sample a wide swath of their surroundings, commonly flying up to three miles from their colonies.

A bee's shopping options at any one time may be limited to a single widely planted crop or broaden in habitats with a patchwork of distinctive microlandscapes and wide floral diversity. Further, a colony's landscape supermarket changes as a variety of flowers bloom through the seasons. Each hive's honey will be different, depending on which flowers its scout foragers discover geographically and seasonally.

The master of honey *terroir* is Minnesota's Brian Fredrickson, whose Ames Farm connects consumers with the floral imprint of each hive and the land from which its honey came. Each jar of his honey tells a unique story, extracted from individual hives every few weeks from spring to fall. Customers can go to his website and match a number on the bottle with the hive, apiary location, and season from which that jar's honey originated.

Fredrickson didn't begin as a beekeeper. His university degree was in engineering, and he worked at 3M for thirteen years before fulfilling his dream of buying some farmland in Minnesota. He took the plunge when five acres of historically important property came up for sale, an orchard where the apple variety Honey Crisp was first planted commercially after its development at the University of Minnesota's Horticultural Research Center in 1991.

Fredrickson believed he was getting into apples, but there were two hives on the property. Before long, bees had eclipsed apples as his passion, and varietal honeys produced from apiaries throughout Minnesota became his trademark.

Unusually, Fredrickson doesn't blend honey from the hundreds of hives he keeps at more than twenty apiary sites, nor does he heat and pump it through industrial-strength equipment. Rather, his artisanal honey is extracted hive by hive, as gently as possible, with the honey-laden combs kept in the "honey house" at the mild colony-like temperature of ninety-five degrees Fahrenheit, but is not otherwise heated. That way, the floral aromatics that give honeys their distinctive tastes are retained, and the honey maintains its natural enzymes and antimicrobial properties.

The warm combs are uncapped with a hot knife and then placed in a round centrifuge-like extractor, each comb in its own individual metal holder, and spun. Centrifugal force drives the honey out of the frames to the inner walls of the extractor, where it runs down and out of a spigot into holding tanks until it's bottled and sealed. The honey is minimally strained on its way through cheesecloth or a similar loosely porous filter to remove wax particles.

The resulting honey is superb—and unique. *Food and Wine* magazine chose Ames Farm's buckwheat honey as the best artisanal dark American honey, calling it "robust and deeply complex; looking and tasting like molasses, with malty, earthy flavors." Dara Moscowitz, a writer for the *Minneapolis City Pages,* a weekly newspaper, experienced Fredrickson's varietal honeys in a state of near ecstasy, eloquently describing their *terroir:* "Here is the taste of two weeks in a particular grove of soaring basswood trees digging their roots into the sheltered floodplains of the Minnesota River Valley. Here is the flavor of a month in the life of a stand of gnarled, wind-twisted horse chestnut trees living by the Blue Earth River. Here is the essence of a scrubby and scrappy stand of red sumac soaking up the sun out by Parley Lake, near the town of St. Boni."

The specificity of Fredrickson's honey extraction creates distinctive varietals, with each honey carrying the unique characteristics of whatever bloom was dominant in the field. Moscowitz dug deep into an epicure's lexicon to describe

Ames Farm honeys: "The elderberry honey is as clean as mist and tastes of limes and gardens in late spring riot. The boneset tastes like autumn pork, roasting in a pan with mint leaves, raspberry tea, and lemon peels. The precious limited-edition buckwheat tastes like an afternoon in an old library, all gingerbread, port, currants, leather, tobacco, and wood smoke."

I talked with Fredrickson last summer by phone about how his customers experience his artisanal honey. I caught up with him in a beekeeper's most common summer habitat, driving in his pickup truck from one apiary to another. Even on the road he was clear about the significance of time, place, and story: "I realized a box of honey was a discrete little snapshot of place. The idea that you put a hive of bees down on some land and they capture the weather there, the plant life, the topography, each a snapshot in time, summer wildflowers or fall blossoms—it's all reflected in the honey. Every person who picks up that jar sees something a little different. They may have a special attachment to some place, a farm history with granddad. It's pretty powerful, really a reflection of people and a place. Some people are just totally blown away that honey can taste so good and be so different, that land had any impact on it at all."

Fredrickson is not alone in connecting place and honey. Virginia Webb runs Mountain Honey in Clarkesville, Georgia, with her husband, Carl, and shares Fredrickson's passion for artisanal honey. Georgia born and bred, loquacious Webb speaks with a soft Southern accent, and she spins a good yarn. Her daddy and granddaddy kept bees, and her accounting and banking background have served her well in managing her artisanal business.

Mountain Honey's focus is local, but its reputation global. It was awarded the gold medal for "Best Honey in the World" at the first-ever World Honey Show in Dublin, Ireland, in 2005 and again four years later at the World Beekeeping Congress in Montpellier, France. Apparently the word has spread: Mountain Honey recently shipped an order of honey to the Palmer weather station in Antarctica.

The Webbs market two premier honeys, one wildflower and one a specific varietal, sourwood. Virginia calls the wildflower "a botanical plate of florals." The light amber honey comes from spring flowers, a mix of nectars primarily from tulip poplar, blackberry, basswood, black locust, maple, and wild blueberries. Each year's product is subtly different, depending on which flowers bloom most abundantly in the spring.

They harvest wildflower honey around the end of June, making room in the hives for the summer sourwood flow. The forty- to sixty-foot-tall sourwood tree, also called Lily of the Valley, has clusters of white, bell-shaped flowers that are a signature of the mountain landscape of north Georgia and western North Carolina.

Webb and her customers have a particular love affair with sourwood honey, connecting it strongly to those ancient Appalachian mountains and the tree from which it comes: "We specialize in sourwood, a varietal honey. People love the texture of it. It's very light in flavor, not a heavy bouquet. Honey's the soul of the flower itself. When I eat sourwood honey, I want to taste that flower, taste that goodness that the flower itself is giving to me."

Conversing with Virginia is itself an artisanal experience, her comments peppered with her passion for the connections between the land, the beekeeper, the honey, and the customer. Mountain Honey stays small and hands-on, running only four hundred colonies at a time without chemicals and antibiotics. She explained why locale and excellence are so important in her relationship to bees and honey: "I think that customers look for quality. Why does anyone buy a bottle of Rothschild rather than, you know, a bottle of rotgut? You want to know the quality of the person who made it, of the bees.

How do you know the quality of the honey if you don't know the beekeeper?"

o o o

Clearly, artisanal honey reflects our deep-seated desire to connect personally with the farmers who produce our food and the land from which it comes. But there also is a dark underbelly to honey, with intrigue that would fit comfortably in an international crime thriller.

This shady worldwide trade in honey thrives by means of a complex and shifting corporate web of producers, shippers, and packers. Hiding behind a maze of obfuscating paperwork and shell companies that span the planet, it provides a window into the murky side of global food commerce.

And there is big money involved in this epidemic of mislabeling, toxic residues, tariff avoidance, and adulteration. In September 2010 the US Food and Drug Administration announced indictments against eleven German and Chinese executives, six companies, and their US representatives, charging them with avoiding $80 million in honey tariffs and selling honey tainted with banned antibiotics. According to the *Globe and Mail*, a Canadian newspaper, this was "the largest food fraud in U.S. history."

The scam was elaborate. The cost to a Chinese beekeeper to produce a pound of honey is around US$0.25, while the same pound costs an American beekeeper about $1.25. To even the playing field, Chinese honey has an added tariff of about $1.00 a pound. To avoid the tariff, honey from China was labeled as originating in countries that pay little or no tariff, including Russia, India, Indonesia, Malaysia, Mongolia, the Philippines, South Korea, Taiwan, and Thailand. The shipping paperwork was altered to indicate that those were the countries of origin. In addition, the drums containing the illicit honey were often mislabeled as molasses or sugar syrup to avoid raising concerns about how much honey was entering the United States.

In total, 606 shipments of Chinese-origin honey illegally entered the United States between March 2002 and April 2008. Much of this honey was contaminated by chloramphenicol, a broad-spectrum antibiotic used to treat serious infections in humans but not approved for use in beekeeping. Presumably the Chinese beekeepers were using chloramphenicol to combat bacterial diseases resistant to the antibiotics that are licensed for use in bee colonies.

Honey samples were tested in a German laboratory and, if found to contain the rogue antibiotic, were shipped through other countries to reduce suspicion. Loads that were discovered and rejected at US ports were relabeled and once more rerouted through other countries and thus managed to evade detection when they again landed in the United States. Other shipments not covered under this indictment were contaminated with chlordimeform, a miticide banned in most countries because it causes urinary bladder cancer. Again, dirty shipments were turned back, then made their way to other ports after being relabeled.

Even the antiterrorist US Department of Homeland Security weighed in, concerned about threats to the US food supply. Gary J. Hartwig, special agent in charge of Immigration and Customs Enforcement, said that this "international fraud conspiracy engaged in illegal and predatory trade practices threatened our nation's domestic honey industry. The crime of importing mislabeled and adulterated goods restricts U.S. competitiveness in domestic and world markets, creates an uneven playing field for American businesses and honey importers and packers who play by the rules . . . and violate[s] the laws and regulations that are put in place to protect U.S. businesses and the American public."

The indictments resulted in five guilty pleas and sentences of eighteen to thirty months of jail time. Indictments against the remaining individuals in Germany and China have yet to go to court. Two of the largest honey-processing companies in the United States, long suspected by beekeepers of importing and adulterating illegal honey, agreed to pay $3 million in fines and are currently embroiled in a class-action lawsuit brought on behalf of all commercial beekeeping operations in the nation.

Sadly, this case was not an isolated incident. Journalist Andrew Schneider reported in *Food Safety News* that millions of pounds of Indian honey banned in the twenty-seven countries of the European Union and elsewhere because of lead and antibiotic contamination were imported and sold in the United States. Importation records show that 45 million pounds of Indian honey—an amount that far exceeds India's capacity for honey production—entered the United States. Schneider speculated that this honey actually originated in China. Ironically, India likely supplied China with chloramphenicol, the antibiotic that led to the ban.

The trade in illegal and contaminated honey is not just the work of deceptive offshore trading partners; US honey packers, including the two major packers who paid large fines, are complicit. Richard Adee, a South Dakota–based beekeeper who manages eighty thousand colonies, didn't pull any punches when asked about the role of American honey packers: "Some of the largest and most long-established U.S. honey packers are knowingly buying mislabeled, transshipped, or possibly altered honey so they can sell it cheaper than those companies who demand safety, quality, and rigorously inspected honey."

Commercial beekeepers generally sell their honey in drums to packers who process, bottle, label, and market distinctive brands that are an amalgamation of many beekeepers' products. Packers have long been a suspect "species" for beekeepers, and from whispered hallway conversations at beekeeping meetings it appears that honey adulteration is rampant. An article in a 2011 issue of the *Journal of Food Science* ranked honey third in the list of adulterated foods, just behind olive oil and milk.

Unscrupulous packers add high-fructose (HF) corn syrup, malt sweeteners, rice syrup, brown sugar, or barley malt to honey. Particularly common is HF-corn syrup, which is produced by industrially converting the glucose in corn to fructose, thereby making it taste sweeter, and costs only one-fifth the price of honey. Its ubiquitous presence as a low-cost filler in many food products that previously did not contain added sugars is perhaps the single most significant contributor to obesity in North America.

With regard to the extent of honey adulteration, a 2011 survey by *Food Safety News* showed that 75 percent of honey on store shelves had no pollen in it. All honey has at least a few grains of pollen that remain in it after normal straining, and that pollen is the only definitive way to determine country and even region of origin. A complete lack of pollen indicates one of two things: The jar has no honey in it at all, or the honey has been ultrafiltered by heating and forcing the honey through tiny filters to remove all of the pollen.

Regular filtering is sufficient to remove wax and bee parts, and, as Richard Adee explains in *Food Safety News,* "There is only one reason to ultra-filter honey and there's nothing good about it. It's no secret to anyone in the business that the only reason all the pollen is filtered out is to hide where it initially came from. Honey has been valued by millions for centuries for its flavor and nutritional value and that is precisely what is completely removed by the ultra-filtration process."

Adulterated honey can still be detected from its sugar profile, but that requires some sophisticated tests that few laboratories are capable of conducting. And while honey is important to beekeepers and honey lovers, it's not enough of a big-ticket food item to get the regulators overly interested in conducting a mass sampling of supermarket honeys. Ironically, adulterated honey was one of the major concerns that ultimately led to the 1906 US Food and Drug Act, which in turn resulted in the creation of the Food and Drug Administration (FDA).

Beekeepers tend to become very upset when they talk about the lack of government support for beekeeping, but even by their standards, reactions to the FDA are extreme. Beekeepers in the United States can't understand why their federal government isn't doing more to protect the industry from adulterated honey and tariff-avoiding shipments from overseas.

I asked Virginia Webb of Mountain Honey why she thought the FDA was so passive with regard to honey issues. She saw the matter in economic terms but also echoed Homeland Security in wrapping honey purity in the American flag: "It's the FDA's job to protect the consumer from fraudulent and misrepresented products and also the FDA's job to protect honest beekeepers from being undercut economically by this unscrupulous adulteration of honey. It's an economic disaster for the American beekeeping industry. We can't be a free and prosperous country if our food source is going to be controlled by someone else."

∘ ∘ ∘

After spending too much time thinking about dirty honey, I sought a cleansing antidote and found it at the Eastern Apicultural Society (EAS) honey show. It's a fierce competition, with the winner receiving a blue ribbon and bragging rights to the best honey in eastern North America.

The EAS attracts more than seven hundred beekeepers every August to an eastern American or Canadian university.

They sleep in dormitories and take over lecture halls, student unions, laboratories, and classrooms to eat, breathe, and talk bees. The weeklong gathering includes close to a hundred lectures, hands-on demonstrations in the bee yard, social events, a honey-trading room where you can drop off a jar of your honey and exchange it for someone else's, and a spirited show where beekeepers compete for the gold in many bee-related categories. The 2013 EAS was at West Chester University outside of Philadelphia and took place in a roped-off area in the student union. The show is not only about honey but also beeswax, baked goods, mead, sewing items, photos, gadgets, and miscellaneous arts and crafts.

Honey, though, is the premier item, with liquid extracted honey divided into white, light amber, amber, and dark color categories, as well as competitions for chunk honey in comb and crystallized honey. The objective is not only to add a layer of friendly competition to the meeting (and beekeepers can be very competitive about their honey), but also to use the show as an educational tool for beekeepers to learn how to prepare a quality product for sale. The judging emphasizes characteristics that separate superior from inferior honey, and honey as a natural, relatively unprocessed product.

I talked with one of the judges, Bob Wellemeyer, who speaks with a gentle drawl that places him geographically as a Virginian. He sports a well-trimmed gray beard and speaks hesitantly but with a quiet passion about bees and honey.

He's a member of a small cadre of judges comprising a handful of highly trained beekeepers who learn the judging craft through rigorous courses and apprenticeships. A typical judge will begin at a community level, judging at county fairs or annual meetings of local clubs, before moving up to the state fair level and eventually being selected for regional events such as EAS.

Once the scrutiny begins, the judges wear white gloves and carefully measure or observe the qualities of each of the three required jars for an entry. Wellemeyer described for me generally what they are looking for: "When you take honey from the hive, it's in its most perfect state. Everything we're looking for is an adverse result of processing."

Judging is based on subtle deviations from perfection. The worst transgression is for the honey to have a water content that is either too high or too low as measured with a refractometer, a handheld device that determines the amount of water in honey. A water content that is judged to be too high means the honey was taken from the hive before the worker bees could evaporate the water below the 18.6 percent level required to prevent fermentation. If too low, below 15.5 percent water, the beekeeper has probably heated the honey to move it more quickly from hive to bottle, thereby damaging the aromatic compounds and subtly changing the honey's distinctive flavor and aroma.

The judges also use a polariscope, a wooden box with glass filters that passes light through honey and highlights imperfections such as lint from the filters through which honey is passed before bottling, air bubbles, crystals, dirt, wax flakes, and foam. They use a simpler tool, toothpicks, to taste the honey to determine whether it has been burned or has picked up a metallic flavor from extracting equipment. Finally, they pay a lot of attention to jars, looking for fingerprints, consistent fill level, and dirt inside the lid.

It's serious business, and judging is particularly stringent at a high-level show like the one the EAS sponsors. As one judge put it, "we have to hit them hard" in order to identify the subtle differences that separate a blue-ribbon honey from an also-ran. From their quiet concentration and generally silent scrutiny, it was clear that the team of judges I observed had worked together before; their concentration was broken only occasionally with comments like "This jar is dirty as all get-out" or "This jar looks real good on crystals."

Mountain Honey's Virginia Webb also judges honey at high-level shows, although not at the 2013 EAS. As she is an international award winner as well as a judge, I asked her why these shows are important: "Honey shows let you compare your honey with others. When you see the quality of honey that others may be doing, it's something you can strive for. I want every jar of honey that I make to look like a winner. When people open that jar, they're going to know that the quality is the same as my show honey. I want it to look like a winning jar of honey, I want it to taste like a winning jar of honey. There's a lot of pride around doing honey shows. I hope that people who enter honey shows will not do it for the ribbon but to improve the way they present honey and the way they think about the customers."

o o o

I never entered a honey show, but I do have honey credentials. For close to thirty years we produced "Heavenly Honey" at the university where I was teaching, selling more than ten thousand pounds a year from up to two hundred colonies. Our primary product was research data, but I'm a pretty fair beekeeper, and honey sales generated enough income to support a few scholarships each year.

The experience of keeping bees and giving talks to beekeepers about their research proved invaluable for my students. It was a source of pride that we could hold our own in bear-pit conversations among beekeepers, sharing stories and bragging about our honey crops. We argued passionately about whether cow patties or burlap worked better in our smokers, whether to use plastic or natural beeswax foundation in starting comb, and what kind of paint lasted longest on hive boxes.

Honey was the connector, engaging us with landowners who hosted our bees on one end and customers who purchased our honey on the other. Heavenly Honey also linked us to the earth from which it flowed, with the land's rich store of natural and human history.

Our apiaries were out in the Fraser Valley, a gently rolling floodplain where the majestic Fraser River emerges from British Columbia's high mountains and steep canyons and flows to the complex estuaries south of Vancouver, then out to the Straits of Georgia and the Pacific Ocean. It's a narrow, fertile valley about sixty miles long, with pastures supporting a thriving dairy industry up against the mountains and extensive raspberry, blueberry, and cranberry fields and market gardens farther along in the western valley, closer to Vancouver.

We were tutored in valley history, natural and human, by the honey it produced and the residents we encountered. We had our favorite apiaries, each with its own distinctive geological and floral character as well as links to the early settlers from the past and quirky human inhabitants in the present.

My favorite apiary was Florence's, a bee yard we inherited from Henry Barten when he retired as the Fraser Valley's government bee inspector. Henry was Dutch, as were many of the valley's settlers, an excellent beekeeper, and the most helpful inspector you could imagine until you violated British Columbia's beekeeping regulations. The apiary at Florence's had been passed from best beekeeper in the valley to best beekeeper in the valley for close to a hundred years, and as a young researcher just beginning my career I considered it a deep honor to have it passed on to us.

We kept our bees up on a ridge near the small rural community of Mount Lehman, a village established in 1864 as a paddle-wheel steamboat pier. The site was dense with towering cedar and fir trees at that time but had been logged by the Lehman family long ago and cleared for agriculture.

Florence herself had lived there for most of her life. She was in her eighties when we met, having raised her family in the ramshackle house she still inhabited. Now she lived on her own, husband deceased and children long gone, with an old black Labrador retriever and a cow she kept in the pasture next to our apiary. A stop to check in with Florence was mandatory on each visit. Somehow she always seemed to know we were coming and had tea ready with her freshly baked peanut butter cookies.

Our bees were shaded under gnarled and spindly apple trees that, though abandoned for some time, had once been part of a producing orchard. Nearby were chicken and llama farms, but more important for the bees were the blueberry and raspberry fields, vegetable gardens, and a considerable amount of meadowland that provided dandelion, clover, fireweed, and other wildflower nectar sources for the bees. Maple trees dotted the landscape, providing the season's first serious nectar each April.

Much of our Heavenly Honey was produced here, known for the distinctive flavor that emerged from Florence's blend of crops, wildflowers, and trees and for its faint greenish tinge, which came from that early spring maple flow. We used the apiary for more than a decade but eventually moved on when Florence passed away and the acreage was sold for a supersized estate home.

The character of much of the Fraser Valley was changing, with swimming-pooled mansions replacing the small farms, just as the farms had replaced the temperate rain forest decades earlier. Diminishing honey crops were a sad indicator that Florence's bee yard was no longer hospitable to honeybees.

Our apiaries often followed the Fraser River. One site, Shangri La, was nestled in a crook of the Fraser, with a stunning view of Mount Baker in the background and bright yellow clusters of yellow broom providing a magnificent source of pollen as our colonies grew in the spring. In nearby Bradner we had Smirle's, a daffodil farm where one of my students had grown up and where his mother still lived. Thousands of Vancouver city dwellers descended on a spring weekend each year for the annual daffodil festival, providing an opportunity for us to talk bees and flowers with the visitors.

Another Bradner apiary, owned by Wayne and Elyse Gates, was just down the road, close by the general store, where my students would stop in for the latest local gossip. The store's signature "Screamers," a mix of slushy and ice cream, was the perfect antidote for a too-hot summer day in the apiary.

Gravel Pit apiary recalled a personal memory each time I visited. It was another inherited bee yard, this one passed on from the British Columbia Ministry of Agriculture when budget cuts forced it to abandon its demonstration hives. This was indeed an abandoned gravel pit, sheltered from the wind and close to hazel-nut plantations that provided pollen for the bees as early as mid-February.

It was a poignant site for me, as I had brought my daughter there one spring when she was three with high expectations for her first encounter with bees. Her name, Devora, means "bee" in Hebrew, and in naming her I had hoped that she, too, would come to love bees as I did. We walked into the apiary, and within minutes she was stung on the ear, beginning a lifelong aversion to all things insect. Visits to Gravel Pit reminded me of how easily our too-defined parental ambitions for our children dissolve as their own personas emerge.

The landowners loved having the bees and the free buckets of honey we paid as rent each year, connecting them to their own land and providing an identity as the beekeeping landlords in their community. Our bees informed them about what was growing on their own and nearby property and evoked reflections about what had been there before.

Sandwiches and cookies, as well as coffee on a cold day and cold drinks when it was hot, were shared along with conversation about whatever fascinating things our bees were doing that day. Chats about bees were accented with local politics, the latest antics of strange neighbors, and the growing importance of conserving natural and farmed land as urban sprawl snaked its tendrils into rural life.

Our apiaries became local landmarks, inspiring directions to "go about three minutes along such and such road and turn left at the first intersection after the bees." The owners' trips to the local stores were punctuated with their neighbors' queries about how the bees were doing and when the honey would be ready for sale.

Heavenly Honey customers completed the connection from the landowner through my laboratory to honey lovers. And our customers were devoted; Heavenly Honey often sold out within days of bottling. One woman from Taiwan bought hundreds of pounds each year and sent the jars home to her relatives, who preferred our honey to their local brands. Another customer—a Russian—had been exposed to nuclear fallout during the debacle at Chernobyl. His mother, a naturopathic physician, believed that our honey's healthful properties augmented her son's diminished immune system. We also traded buckets to a university faculty member who returned it to us in the form of mead months later, and we developed a robust market in the Jewish community, who used Heavenly Honey for the fall Rosh Hashanah holiday tradition of dipping apples into honey.

But it was the honey we gave away to friends, colleagues, visitors, and anyone at the university who had helped us over the past year that was the most affecting. We presented honey to the university's board of governors and to my department's secretaries, to the staff who cleaned our laboratory and offices, to the deans who smoothed our administrative paths, to the cheerful women with Italian accents who made specialty sandwiches for my cafeteria lunches, and to university presidents who came and went every five or ten years.

We also made sure to drop off some cases at the local food bank, and guest lecturers in my courses were always thanked at the end of class with a jar. I traded my barber four jars of honey for a haircut and bartered honey with a salmon-fishing neighbor when sockeye were running up the Fraser River. Visiting beekeepers left me a jar of their own specialty honey and took home a jar of Heavenly in return. Each sold, giveaway, or exchanged jar was an opportunity for conversation, a chance to reflect on how reliant we still are on the natural world around us and how vital it is to preserve nature.

A jar of Heavenly Honey carried with it our gratitude to the bees for collaborating on the harvest, the flowers that yielded the nectar, the land itself, which provided a sense of physical place, and the human personalities past and present inhabiting that land.

The core of what I learned from my thirty years of harvesting, bottling, selling, and giving away honey is this: Food at its best carries memories and reflections that go beyond sustenance to connect the personalities who harvest and the land from which they gather, making holy the simple act of eating.

3

Killer Bees

Concern for the supply of our much-loved sweetener honey is one reason the global collapse of honeybee colonies has been in the news for almost a decade, but bees have made news before. The last time they achieved celebrity status, in the 1970s, I had a front-row seat from which I witnessed the introduction and spread of African "killer" bees in the Americas.

The migrating African honeybees had just moved northward from Brazil into French Guiana, a sparsely populated and remote protectorate of France located on the northeast coast of South America. Massive stinging attacks by these highly aggressive bees had caused numerous fatalities, and the bees had disrupted beekeeping everywhere they had spread.

To fight back, the "killer bee team," a professor and three graduate students from the University of Kansas, myself included, had been sent down in 1976 to find out why these bees—introduced from Africa to Brazil twenty years earlier—had been so successful in the wild. We were further tasked with determining whether the bees could be managed by bee-keepers and to advise the US Department of Agriculture (USDA) on whether the bees could be stopped from spreading to the United States and, if so, how.

We and the African bees made our way to French Guiana as the result of an ill-conceived plan to breed a better honey-bee for tropical new-world climates. Honeybees are not native to North or South America, and the European-derived bees imported to South America by settlers didn't do well under tropical conditions. African bees were rumored to be better honey producers, although they had a reputation for also being highly aggressive.

Brazilian beekeeping officials conceived the idea of crossing the African and the European honeybees to produce a gentle and productive hybrid for the tropics. Apparently it hadn't occurred to them that such hybrid bees hadn't resulted from similar crosses in Africa. They called on one of Brazil's foremost geneticists, Warwick Kerr, to bring some queen bees over from Africa and mate their offspring with the local Brazilian European-derived bees.

Kerr achieved considerable prominence during his career, including two terms as president of the Brazilian Academy of Science, and he was the first Brazilian elected to the US National Academy of Sciences. He was also politically active, a lifetime socialist who assisted dissidents in going into hiding or leaving Brazil during the 1964–1979 military dictatorship. Jailed twice for speaking out against torture, he was a hero to many Brazilians.

Kerr's scientific reputation was based on his basic work in bee genetics, but he was also interested in developing better crops and livestock. He worked on projects to increase the vitamin A content of lettuce and to breed more productive stingless bees, tropical bees related to honeybees that also store nectar.

In 1956 Kerr received Brazil's national award for genetics, and used the prize money to finance a research trip to Africa to study stingless bees. In a 1991 *New Yorker* article by Wallace White, Kerr described what happened next: "Some people at the [Brazilian] Ministry of Agriculture asked me to do something for them. They had heard of the high productivity of the [honey] bees in southern Africa, and they asked me to import some queen bees from Angola, South Africa, and Tanzania, hoping to use them in improving the bees we had in Brazil."

Kerr recognized that, while the bees might be productive, they were also highly aggressive. He arranged to quarantine the seventy-nine queens he had imported while the Brazilian research team selected for the productivity genes and against the aggressive behavior. Unfortunately a visiting beekeeper decided he would help out by removing the quarantined colony's queen excluders, entrance screens that are large enough for worker bees to squeeze through but too small for the queen. Twenty-six queens escaped in swarms and established the nucleus of a feral population.

From that humble beginning emerged one of the most successful invasive species in a long litany of human-assisted biological invasions. The feral bees spread at a blistering rate of a few hundred miles per year, colonizing all of tropical and subtropical South and Central America and a good part of the southern United States since their 1990 arrival in Texas.

The "killer bee" moniker arose from occasional situations in which colonies erupted in terrifying attacks. The worst stinging incident on record occurred in Costa Rica in 1986, when a botany student from the University of Miami climbed over a rock and encountered a large, exposed nest of African bees. The bees exploded; he panicked and fell into a crevice. Within a few minutes he was stung more than eight thousand times and died. This type of fatality—in which death is caused by massive amounts of bee venom—is different from the more typical allergic reaction, which may be caused by only one sting in those with a severe bee-venom allergy.

How common are such fatal stinging incidents? A reasonable guess based on incomplete statistics from Latin America is that five to ten thousand fatalities have occurred since 1956, but most of those took place in the early years. Fatality rates dropped dramatically once residents got used to avoiding bees, beekeepers moved their bees away from humans, and public health authorities developed advertising and the proactive removal of potentially dangerous bee colonies. In the United States there currently are only one or two fatalities a year due to massive stinging.

Kerr himself was deeply troubled by the problems caused by African bees: "At the beginning of all this, when there were so many stinging incidents, and it was becoming clear how great a threat African bees were—I got very, very depressed. It was very difficult to cope with the fact that people were dying because of those bees."

When White asked him whether he would do anything differently if he had the chance, Kerr said, "I would leave those African bees where I found them."

o   o   o

Honeybees are just one of tens of thousands of introduced species globally. Approximately fifty thousand foreign species have been brought to the United States alone since the first European settlers arrived, and perhaps even earlier by the first human migrants to the New World thousands of years ago. Many of these species are considered beneficial, particularly crops and livestock, while others have caused devastating economic or environmental repercussions.

Whether helpful or harmful, we humans have terraformed the globe with human-created habitats that barely resemble their natural origins. Foreign introductions have happened in diverse ways: by accidental transport, deliberate importations by well-intentioned scientists, casual transport by curious farmers interested in trying a new crop or variety, and illegal importations of pets or plants by fanciers or hobbyists. Introduced species may bring with them economic gain or loss, but introductions, whether beneficial or harmful, can disrupt ecosystems for decades or even permanently. Eventually we either become accustomed to the intruder or develop ongoing and costly management regimes to keep the invaders in check.

An astounding 98 percent of all US food crops and livestock are introduced, including corn, wheat, rice, cattle, poultry, and many others, according to a 2005 study led by David Pimentel of Cornell University. Iconic images of American corn and wheat fields waving in gentle summer winds or cattle on the range are the products of landscape change, contemporary pictures posed over hundreds of years as the original forest, prairie, and range habitats were deliberately transformed to farm and pasture.

Few observers would say they regret those importations that have provided so much economic value. In fact, few of us even realize today that most of the plants and animals on our farms are not native. Yet the impact of non-native species on native habitats has been incalculable. With preintroduction surveys rarely available, the loss of local biodiversity caused by these introduced competitors is truly beyond measurement. Still, biodiversity experts consider foreign species to be, if not the most important cause of native species loss, then second only to habitat destruction.

In that same 2005 study, introduced pest organisms were shown to be the cause of an estimated $120 billion per year in control costs, as well as diminished crop yields and ecological damage. Weeds are by far the worst introduced pests, causing $27 billion in losses in the United States annually through reduced crop yields and management costs. Rats are next, with

$19 billion in control costs, loss of stored grains, and fire damage from gnawing on electrical wires.

Domestic cats, whose origin is thought to be in Egypt or elsewhere in Africa, are surprisingly in third place. Their impact is assessed at $17 billion annually because of the 240 million birds they kill each year. After cats, the most damaging introductions are pest insects, livestock diseases, ubiquitous pigeons, and imported human diseases, all of which cost an estimated $7.5 billion in 2005.

Deliberate introductions with good intentions, like the African bees, have had particularly egregious effects in Australia. Rabbits, for example, were brought over from England in 1788 to be raised for food. The feral population began with only twenty-four escapees, and their spread in speed and numbers was the most rapid ever recorded prior to that of the African bees in South America. Rabbits have been the most significant cause of biodiversity loss on the Australian continent; they gnaw and kill small trees and native plants, leading to erosion and wholesale destruction of extensive native habitats.

Weeds have been another well-intentioned Australian disaster. About half of 220 invading weed species were deliberate introductions, mostly as garden ornamentals. Today, control of invasive weeds costs Australians US$4.2 billion annually.

o   o   o

The economic impact of killer bees was far short of US$1 billion, but the bees certainly caused damage and generated a frenzy of panic. In Brazil, honey production dropped from 8,000 to 5,000 metric tons between 1964 and 1971, and 90 percent of beekeepers in the state of Santa Catarina quit rather than deal with the hostile imports. The bees arrived in

Venezuela in 1976, leading to a precipitous drop not only in honey production, from 1,300 to 78 metric tons annually, but also in the number of beekeepers, with only about 10 percent still operating by 1981.

Similar plunges occurred in each country after the bees arrived, coupled with thousands of stinging incidents and dozens to hundreds of deaths. Soon the US media began reporting on the bees, whipping up a frenzy of concern unparalleled in the history of introduced insects.

The reporting was undeniably colorful. *Time* magazine warned: "Like an insect version of Genghis Khan, the fierce Brazilian bees are coming." The *New York Times* said the bees were "like the monster creations of science fiction." In *National Geographic* the bees were "possessed by rage," while the *Philadelphia Inquirer* conveyed its message in capital letters: "They've already KILLED hundreds of people, STINGING some thousands of times. The slightest jostle is enough to send them into a VICIOUS FRENZY. And now, they are heading this way."

By 1976 something had to be done, and that something turned out to be the *Ghostbuster*-like mission of the killer bee team from Kansas. Funded by the USDA, we were supposed to find the chink in the bees' armor and figure out a way to stop their spread northward or to control them if they did make it all the way to the United States.

There's no doubt that our mission was triggered by sensational media reports and growing panic among beekeepers in the United States, but underlying assumptions about the human role in managing nature subtly influenced our assignment. For one thing, honeybees had been a foreign invader in North and South America for more than three centuries, following introduction by early settlers who considered it their divinely inspired responsibility to tame jungles and forests until the landscape resembled the familiar, domesticated farms and gardens of Europe.

Honeybees were a small part of massively transforming the Americas into a managed version of nature. Indigenous peoples called bees the white man's fly and saw them as a precursor of settlement because swarming bees colonized areas of North America just slightly ahead of the settlers. Both managed and feral honeybee colonies became a significant part of the evolving landscape, providing economic benefits but likely with considerable ecological impact on native bees and the flowers they pollinated.

Turning the Americas into familiar farms and gardens was the early settlers' form of hubris, but the introduction of African bees in the 1950s reflected an era driven by a different hubris, the belief that science and technology could solve all problems. Pesticides were expected to eliminate pest problems, antibiotics would keep us healthy, households would be transformed by appliances and labor-saving devices, and nuclear power promised a clean and energy-rich future.

It was a time of too much emphasis on benefits and not enough on possible consequences. The African bees were brought over in that spirit: a vision of only positive outcomes and manageable risks.

Our arrival in French Guiana in 1976 came at a time when science and technology were losing some of their luster, and the penalties of progress were becoming more visible as a result of various technological disasters from DDT to thalidomide to nuclear power plant meltdowns. The public image of science was shifting with the recognition that scientists could get things wrong. Yet the belief persisted that whatever impacts we humans made on the planet could be managed or fixed.

So we were sent south to fix the killer bee problem.

o o o

We worked out of a one-story, cinder-block apartment in Kourou, French Guiana, a small town surrounded by empty savanna for a few miles along the coast and then jungle for hundreds of miles to the south. FG, as we called it, was sparsely populated, with only about sixty thousand residents, and was best known for the former French penal colony Devil's Island, which lay a few miles offshore from Kourou.

Although FG was home to a hodgepodge of indigenous mixed-race Creoles, Maroons, and Saramakans, who spoke a local patois, it had also attracted expatriates from all over the globe, many clandestine escapees from questionable pasts. The French Foreign Legion had recently established a base there, allegedly to prevent Brazil from invading. Legionnaires often dropped by our apartment and told us the gruesome reasons they needed the new identities the legion provided after five years of service.

We often ate at an open-air restaurant down the road in Sinnamary that served jungle meat, including jaguar, capybara (a giant rat), alligator, anaconda, and monkey brains. The proprietor was an American, a former bar owner from Florida who had run afoul of some bad people and had made a quick departure after his bar was destroyed by a firebomb. A smattering of Hmong and Vietnamese refugees had recently arrived, escaping the conflict in Indochina, and were taking over the local shrimp industry.

Kourou was also home to the Guiana Space Center, the main spaceport of the French and European Space Agency, located there because rockets required less fuel to launch due to the earth's increased centrifugal force closer to the equator. Because of the sophisticated tastes of the European scientists and staff, we could head out to the savannas for fieldwork every morning and return in the afternoon to fresh baguettes and croissants from the local bakery. Excellent cheese, fruit, pâté, wine, and even fresh-cut flowers were flown in twice a week by jumbo jet from Paris.

We expected to join the other clandestine French Guiana residents and spend the next few years working in obscurity, studying the bees at isolated sites in the jungle and savanna before emerging triumphantly with some answers from our tropical seclusion. What we did not expect was a bizarre episode that landed us on the cover of *Rolling Stone* magazine in July of 1977 and brought an invitation to partner in a can't-lose business scheme to sell "killer bee honey."

The author of the *Rolling Stone* cover story, Ed Zuckerman, was a freelance writer for both mainstream and alternative media, including the *New York Times Magazine, Harper's,* the *New Republic, Mother Jones,* and *Rolling Stone.* Ed specialized in the eclectic, and his portfolio included profiles of the king of Albania and a talking chimp in Oklahoma. So a killer bee assignment didn't seem at all unusual in his repertoire.

He spent about a week with us, visiting the bees and then writing an article flavored with dramatic stinging stories, wry humor, and glimpses into our quirky lifestyle in French Guiana. Ed, like most journalists we met who wrote about killer bees, favored hyperbole: his cover story's tagline was that he had risked "his life . . . to learn the truth about the killer bees."

We shared quite a few personal stories on our long drives to and from the bee yards, and it was clear that Ed was tiring of the freelance writer's precarious economic circumstances. He was an entrepreneur at heart, always seeking more lucrative opportunities during his reporting trips. One day an idea that he felt was just his cup of tea presented itself to Ed.

He was at home with us after a hard day of fieldwork. He decided to stir into his tea a spoonful of some killer bee honey that we had just brought back from the field. Tropical honeys are thinner and sweeter than honey from temperate zones, and, seeing Ed's enthusiasm for the flavor, we jokingly suggested he market it as "dread killer bee honey" and make a killing. He joked back that we should all partner on it.

We didn't think any more about it after that—until more than a year later when friends told us they had seen Killer Bee

Honey in department stores and specialty gift shops. Ed and a couple of investors had bought some honey in Brazil and put together a novelty product that he hoped would finally get him off the freelance treadmill: Killer Bee Honey.

Billed as the next pet rock, it came in a small bottle accompanied by a booklet and an exorbitant price tag. The booklet said things like "As you taste this honey, remember the lives it has cost. And then enjoy it. If you can." To boost Valentine's Day sales, Ed dressed up in a beekeeper's suit and veil and visited stores he hoped would respond favorably to his cheesy campaign sign: "Give Your Honey Some Killer Bee Honey."

Food critics did not respond the way Ed hoped. One wrote that the honey had "the taste of molasses and silage or hay in a country barn." And so Ed and his investors lost quite a bit of money. A couple of years later their story earned the dubious honor of being featured in a book about why small businesses fail.

Yet Ed's story did have a happy ending. He eventually stopped writing about killer bees when a well-connected friend offered him a chance to write an episode for the television show *Miami Vice*. The episode was based on a story about bull semen Ed had written for the *New York Times Magazine*. Not long afterward Ed moved to Hollywood, where he reinvented himself as a producer, writer, and story editor on the megahit *Law and Order*.

Even after finally achieving immense success, Ed apparently has not given up on killer bee honey. The company is still registered in Massachusetts, thirty-five years later.

o   o   o

Our work was unusual for bee research at that time, which tended to consider bees as a managed species in isolation of their habitats in the wild. We approached our studies from a different perspective, probing how the African bees' characteristics could explain their extraordinary success as colonizers. It soon became apparent that these bees were not going to be stopped or controlled, and our default position became one of admiration for their marvelous adaptations.

The African bees were ideally preadapted for tropical and subtropical New World habitats. Honeybees reproduce by colony division or swarming, in which some of the bees and the old queen fly off to a new nest site and leave behind the remaining bees and a new queen. Feral European honeybees swarm only once a year, with relatively large swarms in the spring, and require a sizable insulated cavity such as a hollow tree in order to build a large enough colony to store sufficient honey to survive the winter.

In contrast, the African bees were constantly swarming, producing many small swarms during each episode. We calculated that the French Guiana bees were multiplying by an astounding sixty-nine colonies per year, an unprecedented growth rate for any species, let alone a highly social bee. A conservative calculation based on a relatively low-end average density of six colonies per square kilometer suggests that approximately 150 to 200 million colonies of feral African bees are currently inhabiting the Americas, about four trillion bees, all derived from those twenty-six original colonizers.

Tropical honeybees also have no requirement to nest inside a winter-proof, well-insulated cavity and will build their comb in small cavities, under eaves, or even in the open, suspended from branches or within dense bushes. The Africans' strategy is to construct small nests, swarm every few months, and produce a greater number of swarms with relatively few workers each.

The tropics do have seasons, but they are defined more by rainfall than temperature. Depending on the locale, either the wet or the dry season will be a dearth period without much in flower. During that time African colonies often abandon their nests and move cross-country for hundreds of miles until they discover a region with better resources. That's an important adaptation in tropical habitats because flowering tends to be patchy. Their ability to migrate long distances, send out scouts to determine whether an area has abundant resources, and then, if it does, start a colony there minimizes colony loss during dearth periods.

The sometimes extremely aggressive behavior of the African bees is also well suited to tropical habitats, where predators like ants, wasps, small mammals, birds, toads, and especially humans attack colonies for their honey and protein-rich brood. It's variable, but these bees can have a hair-trigger temper: thousands of bees will be in the air stinging within seconds, a strong deterrent to vertebrate and insect predators.

The same qualities that made the African bees so successful as invaders also made them unsuitable for beekeeping, at least in their feral form. Frequent swarming diminishes honey production, and having many colonies abscond every year makes it difficult to maintain colony numbers. Heavy stinging requires apiary sites to be situated quite a distance from people and also makes their management by beekeepers almost impossible no matter where the hives are sited.

o o o

We returned from South America with the clear message that the African bees were not going to be stopped. Yet that did not prevent governments from trying. Ideas from ludicrous to impractical were floated, including a gas-fueled flame wall along the Panama Canal, a fifty-mile-wide swath of Nicaragua sprayed weekly and forever with the insecticide malathion, and even a similarly wide radiation belt through the jungle.

More substantive projects were implemented to genetically swamp the bees with massive numbers of the more docile European varieties. Even Warwick Kerr succumbed, spending his own money to distribute eighteen thousand European queens in the hope of overwhelming the African gene pool through mating in the wild, an investment of about $400,000 in today's currency.

His attempt failed, but that didn't stop the USDA from implementing a bee barrier zone in Mexico, an even more massive genetic battle plan that included 39,000 bee colonies, 16,000 traps to catch African male drone bees to prevent their mating, 141,000 bait hives to trap swarms, 1,150 employees, and 220 vehicles. The plan failed for a number of reasons, not the least of which was that, by the time the plan was implemented in 1987, the African bees had already spread north of the Mexican barrier zone.

But our research and Kerr's experience suggested it was going to fail anyway, largely because hybrid bees that result from European-African mating express the dominant African traits. Also, any colonies that did have European-type traits would do poorly in the tropical wild since their temperate-evolved traits are simply not suitable for feral honeybees in tropical habitats.

Killer bees illustrate in microcosm the issues that arise from the deliberate introduction of nonnative species. There's the preintroduction optimism about benefits, followed by a realization of negative environmental and economic impacts, growing public concern due to media coverage, and then the traditional "we don't know enough" scientific response. Next, expensive efforts are implemented to repel the invader, usually unsuccessfully. Failure is followed by adaptation and a new status quo that accepts the presence of a now-unwelcome inhabitant and the ongoing management paradigms needed to minimize its impact.

The history of honeybees in the New World also reflects back to us how our thinking about environmental effects has progressed. It began with the Old Testament command for humans to have dominion over the earth and by the mid-twentieth century had moved to an unfettered belief in science and technology. The last half of the twentieth century was marked by recognition that progress could have negative repercussions, but we were confident that we could fix them.

Today's global ethic continues to shift. The precautionary principle currently in vogue as the safest regulatory practice insists on a high bar, demonstrating conclusively before implementation that a technological innovation, including the introduction of non-native species, is not harmful. Although we're still not that inclined toward caution, the precautionary principle has slowed the onset of genetically modified crops in Europe and has been a major argument in favor of action to at least slow the rate of climate change. Had the precautionary principle been our dominant environmental ethic in the 1950s, it's unlikely that African bees would have been introduced.

Have we made progress? Precaution seems to be an improvement on divinely inspired dominion over the earth or unquestioned faith in science, but at best it has slowed rather than stopped human intervention on our planet. The underlying tension between progress that is narrowly focused on human gain and decisions that more broadly consider the health and well-being of the natural world remains a fundamental issue.

Our team learned that we couldn't stop the African bees or control their spread, but we could adapt. Ultimately, beekeeping survived the killer bees. Decades of selection for gentler African bees rather than African-European hybrids resulted in more tractable bees, at least in managed colonies. The public adjusted to the idea that honeybees can be aggressive, and problematic colonies are quickly removed from proximity to people.

In Brazil, massive clear-cutting of tropical forests and the planting of nectar-producing citrus orchards have given bee-keeping a boost, perhaps not the best ecological solution to the problems caused by African bees but still a boon to bee-keepers. Ask Brazilian beekeepers today, and many now think the importation of African bees was a good idea, although how much of that is rationalization and how much real is difficult to determine.

o o o

Little of our time as America's killer bee team was spent doing interviews with *Rolling Stone*, drinking with Legionnaires, and enjoying fresh, imported pastries with the locals. Day after day in French Guiana, we went out to our apiaries. We spent most of the daylight hours meticulously examining and measuring every conceivable aspect of colony life, from comb area to the amount of brood being reared to adult population, and more. We climbed high up in trees to catch swarms that had alighted on branches and removed wild colonies from their feral nesting sites, using machetes to chop into hollow logs.

This work with African bees revealed just how far we humans have pushed our own systems out of balance. The African bees had been brought over without consideration of consequences. The importers' belief that we could manage nature had trumped caution about potential pitfalls. Government's policy response was to fear the side effects of importation and attempt to stop the bees from spreading, a fix that our research had indicated was not feasible.

In the end, after many decades we've adjusted to African bees, but attitudes about our capacity to manage nature have not changed. Today bees and the flowering plants that depend on them are facing a new threat, colony collapse disorder, a much more serious problem than killer bees ever were.

Like African bees, colony collapse is a human-caused problem, reflecting the dark side of human pride. It's decimating both managed and feral colonies with implications for beekeeping, crop production, and the integrity of unmanaged habitats.

4

A Thousand Little Cuts

Honeybees are dying all over the globe, and this dire fact has severe economic implications for beekeeping and crop production, not to mention what a deep tragedy their decline is for the natural world. A picture of why bees are in trouble has been slowly coalescing from behind the fog of hypotheses about the reason for the bees' demise, with scientists now concluding that the decline is not due to any one factor but rather many interacting causes.

Perhaps most interesting has been the realization that pesticide and diseases, any one not fatal on its own, may be killing bees by acting together. A number of low-level stressors appear to be working in concert to bring about the catastrophic outcomes experienced by colonies in every region where bees are kept.

This developing story about the collapse of bees is particularly relevant for human health, given that we are exposed to hundreds, if not thousands, of low-level toxicants in our food and environment. Each pollutant alone is supposedly safe for us, according to government regulators, but for the first time we are learning—from bees—that the regulatory limits to exposure for single compounds provide little safety.

Here is our conundrum: It's late spring; strawberry season. I just bought a large container of plump berries, perfect in smoothies or with my daily breakfast of organic yogurt, homemade granola, and honey.

But my anticipation of eating pleasure is tempered by my awareness that strawberries are ranked number two on the Dirty Dozen, a popular and widely consulted consumer warning system highlighting the crops most contaminated by pesticides. It's published annually by the US Environmental Working Group (EWG), an influential advocacy organization, and generates a flurry of media attention each year. The EWG reminds us that those tasty strawberries come with residues of many fungus- and insect-killing chemicals, their potential toxicity hidden behind trade names like bifenthrin, boscalid, captan, cypronidil, fenhexamid, fludioxinil, malathion, myclobutanil, pyraclostrobin, and pyrimethanil.

My partner, Lori, and I consume copious quantities of many of the EWG's Dirty Dozen, including apples, the number one most-contaminated crop, as well as celery, spinach, grapes, peppers, blueberries, lettuce, and kale. Each has residue profiles similar to strawberries, although each possesses its own unique pesticide signature.

Toxicologists tell us it would take a million times more of even list-leader apples to show a minimal health effect. Carl Winter and Josh Katz, food scientists at the University of California in Davis, conducted a thorough study of residues in the Dirty Dozen and concluded that: "Exposures to the most commonly detected pesticides on the twelve commodities are at negligible levels . . . and pose negligible risk to consumers."

The EWG itself sends out mixed messages. Its Dirty Dozen report frightens us with "U.S. and international government agencies alike have linked pesticides to nervous system toxicity, cancer, hormone system disruption, and IQ deficits among children." But then the organization dithers: "Eating conventionally grown produce is far better than not eating fruits and vegetables at all . . . Health benefits of a diet rich in fruits and vegetables outweigh the risks of pesticide exposure."

It's almost impossible for consumers to determine whether we're at risk from pesticide residues in our food, and pesticides are just a small part of the toxicants in our environment. For example, PVC building materials expose us to minute amounts of phthalates, which have been linked to childhood asthma, while our backyard decks and picnic tables are treated with arsenic, associated with bladder cancer. Meanwhile, hormones from women's birth control pills end up in their urine, which in turn winds up in our drinking water and may be responsible for the early onset of puberty in young girls. What's more, the vitamin A that's in sunscreen accelerates the emergence of skin cancer rather than prevents it.

Our issue as consumers is that we're exposed to numerous regulator-approved compounds, each of which individually has passed scrutiny to ensure health and environmental safety. But almost none have been evaluated in groups of two, three, ten, or hundreds for their collective effects. How do the ten pesticides in those strawberries interact in my body? Are they additive to or interactive with other pollutants?

Shockingly little research has been conducted on synergy, defined as "combined action being greater than the parts." But one of the research hot spots has come from colony collapse disorder (CCD), the term given to the recent global collapse of honeybees. Great effort has gone into studying why bees are dying. Scientists initially searched for one answer to CCD, perhaps a new disease or a novel pesticide. When nothing appeared, they began examining synergy between multiple factors, a less obvious but more insidious phenomenon.

Pesticides were an obvious element to test, including those used by beekeepers inside hives and those applied by farmers in nearby fields. A pesticide on its own, at low concentrations, may have no impact, but add in a mild disease or parasite, and the combination could be lethal. Effects might also accumulate and cause colony demise over a few weeks or months.

Bees have become frontline research subjects to test whether synergy between minute levels of synthetiç compounds as well as interactions between chemicals and diseases or pests might erupt into unexpected toxic effects. The picture that is emerging suggests that one plus one vastly exceeds two. If so, bees may be the early warning sign that we, too, are susceptible to demise by synergy.

o o o

Honeybee colonies are collapsing around the planet, with bees from one-third of all colonies dying each year. It's been major news, the most extensive coverage of bees since the infamous killer bees in South America became media darlings in the 1970s. Although the threat from killer bees turned out to be hyped, colony collapse is developing as a more serious issue, threatening beekeeping as well as yields from the numerous crops that rely on bees.

Colony collapse disorder burst onto the agricultural scene in 2006, when numerous beekeepers went out to their apiaries that spring and found the bees in many of their colonies had mysteriously vanished. It was something beekeepers had almost never seen before. Colonies would seem vibrant and full of bees one week, and the next week most of the adult workers would be gone, leaving behind capped brood (pupae), stored honey and pollen, and even the queen.

CCD has not abated, posing a continuing and increasing threat to beekeeping as well as the numerous crops that rely on bees. Annual colony losses of 30 to 40 percent are now routine globally, and losses can go as high as 100 percent of colonies for some beekeepers. On Vancouver Island, for example, a major beekeeping region across the Straits of Georgia from my Vancouver home, 90 percent of colonies died over the winter of 2009–2010.

Beekeepers are struggling to rebuild. Currently in the United States, beekeepers are losing close to 850,000 of their 2.6 million colonies each winter, rebuilding these numbers during the spring and summer, and then losing the same number again the next winter. The number of beekeepers has crashed as dramatically as the number of colonies has collapsed, from 210,000 in 2002 to about 120,000 in 2013. Global colony losses run into many millions of colonies annually.

For perspective, imagine if one-third of a more familiar farm animal, say, cows, died of disease each year, including 90 percent of the dairy cows in the Netherlands and 40 percent of the beef cattle in Texas, bankrupting close to half of the world's ranchers and dairy farmers. Now make that a global phenomenon for bees, and you can grasp the dimensions of this disaster for honeybees and everything that depends on them.

The danger of CCD goes well beyond its effect on bees themselves; much of our food supply is jeopardized by reduced honeybee populations. Bees pollinate about one-third of our crops, including fruits, berries, nuts, vegetables, and the forage crops like alfalfa, which are used for hay to feed livestock. Without bees, we would be hard pressed to feed ourselves.

There also is a backdrop to the CCD story, a contributing factor overlooked when we focus on the tragedy of so many colonies dying. Like most agriculture, beekeeping is not what it used to be. The last three to four decades have seen the slow but relentless transformation of beekeeping from a relatively small-scale, pastoral occupation directed by a close feeling for nature's rhythms, with little chemical input and mostly stationary apiaries, to a highly industrialized business where mobility is essential and management is driven by artificial feed, pesticides, and antibiotics. This industrialization of beekeeping created perfect conditions for an epic collapse.

Some US operations manage up to seventy thousand colonies, or at least used to before CCD devastated the industry. Many of these operations are highly migratory. Typically, colonies are wintered in southern states feedlot-style, in apiaries with hundreds or even thousands of colonies side by side on pallets, fed corn sugar by hose to keep them alive until moved by flatbed truck to California in the spring to pollinate almonds. The hives are then moved again a few weeks later to pollinate plums, then on to apples, trucked north and east for summer honey crops, and finally back south for winter.

The transformation of a process based on small-scale, local, and stationary apiaries into a giant mechanized, mobile enterprise was facilitated by high inputs of pesticides and antibiotics applied by beekeepers within colonies to combat pests as well as prevent bacterial and fungal diseases. In addition, heavy feeding with processed corn syrup and artificial protein supplements became the norm, a highly simplistic diet compared to the complex, diverse, and bee-collected constituents found in floral nectar and pollen.

Colony collapse disorder is not occurring in a vacuum but is affected by practices in agriculture outside of beekeeping that also make bee management problematic. Farms today are composed of vast acreages of single-crop fields treated with insecticides that often are toxic to bees, and weed killers eliminate any alternative forage for bees outside of the crop. Fungicides are applied profusely. They were thought to be harmless to bees until some recent studies suggested that they interfere with honeybee immune systems, increasing their susceptibility to disease.

Reduced forage for bees and heavy use of bee-toxic pesticides in today's mass-farming system would be bad enough, but beekeepers also face an increasing array of honeybee pests and diseases that have been introduced or have become more serious due to pesticide and antibiotic resistance. When CCD hit, each of these was investigated as the potential causative agent. The varroa mite was the most obvious suspect. Accidentally introduced from Asia in the mid-1980s, this mite feeds on the blood of both pupae and adults, weakening individual bees and leading to colony death within one to two years unless controlled.

Beekeepers soon found themselves on the same pesticide treadmill as mass agriculture. The initial antivarroa miticides were applied more frequently with longer exposure times and at higher dosages than labels recommended. Predictably the mites acquired resistance and the chemicals lost effectiveness. The same story was repeated with other miticides; now only the harshest pesticides remain, chemicals that have been phased out for other agricultural applications and are allowed to be used in hives only under emergency registrations.

The baneful effect of varroa alone was not a sufficient explanation of colony collapse, and scientists next turned their attention toward diseases, particularly the many viruses that are triggered in bees weakened by varroa. There are dozens of them, and some new ones were discovered as research progressed.

Viruses, fungal diseases such as nosema, and the bacterial disease foulbrood that had become resistant to antibiotics because of overapplication by beekeepers, were all investigated. Although they each contribute to colony mortality, none could be definitively linked to CCD.

Finally, scientists focused on the pesticides and antibiotics used within hives to control mites and disease and externally to kill insects, weeds, and fungi. Researchers addressed interactions and hypothesized that synergy between pesticides, honeybees, and pests, as well as between pesticides themselves, could be behind CCD.

If so, we will not find a magic-bullet fix for CCD but will need a wholesale reconfiguration of the way bees are managed and a redesign of the agricultural tapestry woven by the complex social behavior and pollinating functions of bees.

And if synergy is at work on bees, can humans be far behind?

o o o

The first significant clue to solving the synergy puzzle came from the Department of Entomology at Penn State University. This department has a long tradition of conducting scientific research on bees and running practical extension programs for the state's beekeepers. Pennsylvania was an appropriate home for the first breakthrough, as David Hackenberg, former president of the American Beekeeping Federation and a prominent Pennsylvania beekeeper, was the first to raise the alarm about CCD when he lost 80 percent of his colonies in 2006.

Penn State's landmark research was itself a product of scientific synergy. The collaborators included lead author and Pennsylvania native Chris Mullin, a toxicologist with broad experience in pesticide research but without previous bee experience; Maryann Frazier, a prominent beekeeping extension agent; James Frazier, a former department chair who had previously worked for the DuPont chemical company and has been a significant participant in agricultural policy decisions in the United States; and a number of scientists from the US

Department of Agriculture. Notably, none of the researchers had any history of opposing pesticide use, and all were well respected in the notably conservative US agricultural and agribusiness communities.

Mullin is more of a toxicologist than an entomologist. Like many chemists, his initial interest came from blowing things up in a friend's basement. He enjoyed this so much that he eventually went on to study chemistry at university. He's prone to wearing sweaters and is quiet in conversation, yet his calm, reasonable, and thoughtful manner makes him a powerful speaker.

Mullin's authority comes from being careful and unbiased, without a pet theory to prove. About five years ago he was drawn into bee studies because, as he put it, bees are an "incredible model insect for sampling the chemistry of the environment as they go so widely and range so far to collect the chemistry that's out there."

The study, published in 2010, breathtaking in thoroughness and detail, has meticulousness in method and calm presentation of facts that lend credibility to controversial findings. The researchers found a mind-boggling 121 different pesticides in the wax comb, from miticides to field-applied insecticides, fungicides, and herbicides, many with known toxic effects on bees.

Maryann Frazier, one of the other authors, was stunned by what they found in the comb. She told me: "The chemicals that showed up were not even registered any more for use. The wax is like a fossil record." Mullin was also "horrified by the results. Compared to other commodities, this was the most polluted in regard to frequency and amounts of different chemicals. That's pretty alarming. It's not any one pesticide correlating with decline or health problems; it's the suite of chemistry and the chemical load that's causing the problem."

Mullin and colleagues named this outcome "toxic-house syndrome" and suggested that this chemical soup may well be having detrimental effects on bee behavior and survival. Constant exposure to even low levels of miticides in wax, for example, is a significant risk factor in selecting for varroa resistance. Some of the higher pesticide concentrations found in pollen could be interfering with larval development, meddling with adults' capacity to perform behaviors such as foraging, shortening life span, or killing adult bees even without the other pesticide residues.

Particularly disturbing was that an average of 214 parts per million (ppm) of pesticides were found in stored pollen, the only source of protein for larval and adult bees. Generally, 214 ppm of many of these individual chemicals would cause problems for bees. Mullin points out that "Ten pesticides were found in pollen at greater than one-tenth the bee LD50 level [the amount that will kill 50 percent of bees] indicating that sublethal effects of these toxicants alone are highly likely." But no one knows what the impact of the aggregation of pesticides in comb might be at this collective dose.

Mullin's speculation was confirmed by researchers at Washington State University, who reared worker bees in comb with levels of pesticide residues similar to those reported by Mullin. They found delayed brood development times and higher brood mortality as well as reduced adult worker life spans for those adults that did develop successfully. And in a particularly cruel twist, the authors noted that delayed brood development would be a boon to varroa, which thrives in situations where the mites have more time to feed on brood.

Mullin became interested in CCD because of his interest in systemic insecticides, a new type of pesticide in which seeds are coated before planting, thus depositing minute but effective levels of pesticide in plant tissues as they grow. These seed treatments are considered progressive because they target only insects feeding directly on crops. They avoid ground or aerial spraying that kills nontarget insects indiscriminately, protect-

ing beneficial insects such as bees and the parasites and predators that are natural control agents for pests.

The most widespread seed treatments today use a class of
pesticides called neonicitinoids because of their close chemical relationship to nicotine, a natural plant-produced compound that protects tobacco from insects. Nicotine was one
of the first pesticides applied in modern agriculture, but it is a
harsh chemical with considerable impact on nontarget organisms. The neonicitinoids also are powerful and tough on bees
when sprayed, but the seed treatments were thought to solve
that problem by keeping the chemicals away from beneficial
organisms.

However, tiny amounts of neonicitinoids make their way
into nectar and pollen, generally less than ten parts per billion, amounts thought to be far below a dose that would have
a lethal effect on bees. French beekeepers became alarmed in
1994, when colonies dwindled and died when placed near
flowering sunflowers that had been treated with one of the first
neonic pesticides, imidacloprid. However, varroa resistance to
miticides also became epidemic in France that year, confusing
the issue of what was killing hives.

The direct impact of neonicitinoids on bees remains highly
disputed among scientists despite almost two decades of research that so far has not established a conclusive link between
colony collapse and neonicitinoid seed treatments. What is
clear is that neonicitinoids can have sublethal effects at the
low dosages bees encounter in nectar and pollen, causing memory loss so that bees may not find their way home to their
nests and diminishing the effectiveness of their immune systems. Whether these effects alone are sufficient to cause colonies to dwindle has yet to be definitively established.

The pesticide industry has vigorously defended its neonic
products, insisting that regulatory decisions be science based
and that the science does not support restricting their use. So

far, North American regulators have agreed, continuing to license neonicitinoids. European regulators have not been as compliant with industry lobbyists; the neonics have been banned on crops attractive to bees for a two-year reevaluation period in Europe, beginning December 2013.

One company, Bayer, has been a particularly enthusiastic supporter of neonicitinoids, not surprising considering that its product imidacloprid is the most widely used pesticide in the world. I asked David Fischer, who is responsible for Bayer's risk assessment file for bees in North America, whether beekeepers' concerns about neonicitinoids were justified.

He told me: "There are a lot of beliefs that get ingrained in people's minds, even the minds of people in the scientific community; oftentimes you find there's not a lot of data there. Scientific data, if objectively evaluated, will continue to support these products. Politically? I don't know. We're all used to thinking that if there's an abnormality going on, we look for a chemical first."

Fischer may be right, but suppose it's not a pesticide alone but an interaction between a pesticide and a disease or between multiple pesticides?

o   o   o

Science provides many moments of revelation, but my synergy moment came at an unlikely site, a highway-exit resort in Orlando, Florida, better known for its proximity to Disney World than for scientific breakthroughs.

The Wyndham Orlando Resort was the location for that year's American Beekeeping Federation (ABF) meeting. Like most annual beekeeper meetings, the ABF meeting is low budget, typically held at outlying hotels ringed by highways, malls, and fast-food eateries. The resort's suburban parking lot was turned into a temporary city of beekeepers, all of the spaces filled by flatbed trucks with cute bee logos painted on their doors, and the meeting fueled by weak coffee, pancake breakfasts, and piles of doughnuts at every break. Fruit was hard to find, even in the center of one of America's foremost fruit-growing regions. Fried carbohydrates temporarily displaced my Dirty Dozen pesticide load.

My synergy epiphany occurred during a talk by entomologist Yves LeConte, who was describing research in his French laboratory that had exposed bees to the neonicitinoid pesticide imidacloprid and the bee disease nosema.

LeConte's life in France couldn't be more distant from Orlando highways and super-sized meals. He lives in the small village of Le Thor, a few kilometers outside the walled city of Avignon, in a fifteenth-century house he has lovingly renovated room by room. He bottles his own wine from a nearby vineyard he shares with friends. Fast food is an oxymoron for Yves. Two- or three-hour meals are typical even at work, and I've never seen him eat bread at home that was more than a few hours from the oven of the tiny local bakery around the corner.

His attitudes about bees and research are uniquely French in a way that makes his work on synergy seem almost inevitable. The scientific compass of LeConte and his colleagues points toward promoting balance between humans and nature rather than toward the more North American direction of managing nature to serve human prosperity.

Their study compared unexposed bees, those subjected to sublethal doses of imidacloprid or spores of the gut-infecting nosema disease alone, and bees exposed to both the pesticide and the disease together. Although imidacloprid or nosema had some independent effects, the dually exposed treatments demonstrated synergy, showing considerably more impact than either single exposure. The results included higher bee mortality and reduced resistance to the disease in the dual-exposure treatments.

The impact was both individual and social, with implications for wild bees as well as honeybees. Individual bees exhibited reduced functioning of an enzyme, glucose oxidase, involved in producing hydrogen peroxide, which adult bees secrete in minute amounts into the food they feed larval bees, thereby sterilizing the secretions and preventing disease. The dual exposure reduced the normally antiseptic colony environment, allowing the nosema spores to generate and induce disease.

I talked with Yves after his Orlando talk, and he acknowledged that wild bees such as bumblebees are also susceptible to nosema-caused diseases and exposed to imidacloprid in the field. He was deeply concerned about the wider implications of these results: "We showed that there is a synergistic interaction between both agents, the infectious organism nosema and a chemical, imidacloprid, that is widely used to eliminate insect pests. This can affect all pollinators, not just honeybees."

At least a dozen other studies have confirmed the result for honeybees and bumblebees, finding similar consequences of imidacloprid as well as other pesticides. One, by USDA research scientist Jeff Pettis and colleagues, succinctly summarized the implications with a conclusion unusually opinionated in normally reserved scientific papers: "We believe that subtle interactions between pesticides and pathogens, such as demonstrated here, could be a major contributor to increased mortality of honeybee colonies and other pollinator declines worldwide."

Not only are pesticides being revealed as disease synergists, but they also synergize each other. David Hawthorne and Galen Dively of the University of Maryland published a study with the telling title "Killing Them with Kindness," reporting that the impact of miticides applied by beekeepers is more severe in the presence of antibiotics, which reduce the bees' ability to detoxify and eliminate pesticides. They describe the mix of these supposedly helpful management tools as a "dangerous chemical combination." Antibiotics similarly enhance the damaging effects of neonicitinoids.

Reed Johnson and colleagues from the University of Illinois summarized yet another study showing that two commonly used miticides interact to reduce bees' ability to detoxify pollutants, harming bee and colony health, pointing to "the potential for the observed synergism between [miticides] to harm whole bee colonies and . . . the risk to bees posed by wax contaminated with years of accumulated miticide treatments."

The overall picture emerging from these and many other studies is one of small effects on individual bees that are amplified as they accumulate in the tens of thousands of workers that make up a colony. Each worker bee's function is reduced only slightly by any one exposure, but synergistic interactions from multiple exposures to diseases, the miticides and antibiotics applied by beekeepers, and the agricultural pesticides collected by foraging bees lead to considerably worse outcomes than from any one factor alone.

These accumulated effects have been modeled by a graduate student at Penn State University, Wanyi Zhu. She used life table analysis, a research tool developed in 1662 by John Graunt, an English haberdasher. Graunt applied birth, fertility, life expectancy, and death data to predict population size and age structure. His work led to the insurance industry's actuarial models, provided a key tool for government policy decisions, and influenced businesses to develop population-based marketing strategies. In spite of his working-class background, he was reluctantly elected a fellow of the Royal Society by the scientists of his era.

Zhu's model focuses on the impacts associated with the pesticide and disease synergies identified in previous research. These subtle effects can include slightly reduced nectar and pollen collection due to disoriented foragers failing to return to the hive, shortened life spans in workers who begin to forage a day or two earlier than they normally would, and small increases in larval mortality, which result in a downstream reduction of the adult worker population.

The result of these subtle impacts is to destabilize the age structure that characterizes a healthy colony, disrupting the well-functioning balance of eggs, larvae, pupae, and adults of various ages. The most significant factor may be precocious foraging by adult worker bees, which is caused by some pesticides and diseases that interfere with a hormone that regulates the ages at which bees begin to forage. Early foraging removes younger bees from in-hive tasks such as feeding the young, resulting in higher brood mortality and other cascading effects in a downward spiral of colony decline.

It's as if most adult humans in my town of Vancouver were to wander off over a few weeks or months, leaving babies and toddlers to be tended by teenagers. Some teens would prematurely become responsible caregivers, but it wouldn't be long before our age-shifted human society would collapse into turmoil and dysfunction.

Remarkably, Zhu's model mimics colony collapse disorder almost perfectly. Superficially, all seems well in her modeled colonies, but then they suddenly collapse and die. The slow crunch of small demographic challenges builds until a colony's natural resilience is overwhelmed.

o o o

Perhaps Chris Mullin's term, "toxic-house syndrome," is a better name for colony collapse disorder. Research has demonstrated synergy in honeybees between the effects of a few pesticides and a few bee diseases acting simultaneously, and we've only scratched the surface of interactions. The impact of synergy is tragic and elegant in its simplicity. A pesticide or an antibiotic may reduce a bee's ability to detoxify other toxicants, diminish the immune system's capacity to respond to disease, or limit worker bees' production of antibacterial and antifungal agents that sterilize larval food. Pick one and then add a light case of a disease such as nosema, which induces worker bees to forage at an abnormally young age, or a virus triggered by a varroa mite infestation, and synergy turns deadly.

An individual bee can rebound from one of these afflictions, much as we would shake off a cold, but not two or three disorders. Further, the research has revealed that these multiple effects do not simply add up. Disorders act together and escalate exponentially through synergy to induce considerably more damage together than simple addition would predict. The effects of these simultaneous attacks are writ large on thousands of bees, not only damaging individuals but also disrupting the tightly woven social fabric of the hive. Colony-wide catastrophe is the result.

Bees are dying due to a mix of factors that varies colony to colony, apiary to apiary, and region to region. Colonies may encounter diverse and abundant flowers in one region but face high concentrations of neonicitinoid insecticides in nectar and pollen and have to deal with an outbreak of nosema. In another locale, particularly resistant varroa mites may have evolved due to overapplication of miticides by beekeepers, triggering viral outbreaks, which, when added to high miticide residues in comb and a few relocations for pollination, can do colonies in.

Lessons learned from the synergy-based challenges of honeybees go well beyond beekeeping, into the profoundly complex challenges of regulating chemicals that influence human health and well-being. To date we know almost nothing about how the myriad toxicants that permeate our homes, schools, workplaces, food, fields, forests, and water systems interact. Each toxicant may be well regulated alone, but our exposures are not to isolated chemicals. Rather, we're exposed to a profusion of synthetic compounds in a potentially toxic package of pollutants.

Suppose I'm exposed not only to the ten pesticides on my morning strawberries but also to the arsenic on my deck and sunscreen on my face and arms, and each is present at one-millionth of the dose that would harm me. Do their effects simply add up? If so, the total impact of the twelve compounds would equal an infinitesimal twelve-millionths of a harmful dose. Or are they synergistic, harmless independently but reaching a harmful dose when taken together because of how they act in concert?

Suppose I have the flu or had a mild heart attack last month, or my prostate cancer is in remission following radiation treatments, or I'm nine months old and my brain's development is in overdrive?

Bees have provided us with the first close look at the potential for synergistic interactions between pollutants, diseases, nutritional deficiencies, and other factors. We, too, live in a toxic house in which we are exposed to many combinations of toxicants that interact in almost completely unknown ways with our gender, age, health, and previous exposures., Honeybees are a particularly useful model to probe these potentially harmful human synergies, because they exhibit impact at individual, societal, and environmental levels.

Even analyses of the effects caused by single toxicants have overwhelmed the capacity of regulators to protect us from industrially produced chemicals. Biologist Sandra Steingraber, in her 2011 book, *Raising Elijah,* describes some well-supported associations between childhood diseases and toxic exposures, including preterm birth, which has been linked to exposure to particles and combustion products from coal, and asthma, with its connection to ingredients in plastics. Learning disabilities have been associated with air pollution, organophosphate pesticides, and heavy metals such as lead, mercury, and arsenic. Autism has been linked to chemical exposures during early pregnancy, and premature female breast development has been associated with hormonally active chemicals.

Notably, regulators have had little success in eliminating exposures to these already-convincing singly harmful toxicants. Should synergy between toxicants and other factors emerge as demonstrably harmful, such results would strengthen public pressure for tighter regulation. If research on synergistic interactions does demonstrate threats to humans comparable to the synergistic effects on bees, we will have to prepare ourselves for a more serious commitment to regulation.

Maryann Frazier, collaborator on the Penn State pesticide study, recognizes the regulatory mountain that must be climbed to consider synergy when analyzing potential harm to bees: "Risk assessment today is based on one material at a time, but that's not the reality of what's happening in the fields. What is required is for the Environmental Protection Agency to change its culture in terms of what it requires from registrants. It's such an unwieldy organization, way too supportive of the chemical industry. They know they need to change— but there are so many layers and so much bureaucracy."

Even though ethical considerations prevent the kind of experimentation on humans that we do on bees, epidemiologists and toxicologists have many effective techniques to unravel complex synergies. A good place to start would be to simply explore the effects of two factors simultaneously on vulnerable populations, particularly children, seniors, and those already disabled by disease. There are many mammalian models that stand in for humans in experimental studies, and statistical techniques can be powerful in teasing out synergistic interactions if sample sizes are robust.

There is one body of research examining complex chemical interactions that suggests studying synergistic effects on human health is feasible, and that is the voluminous data on pharmaceutical interactions. Use of multiple prescription drugs is rampant globally, averaging six to nine separate pharmaceuticals for those over fifty-five years of age in the United States. The estimated cost of interactions between drugs that alone are considered beneficial is $130 billion annually due to illness or mortality.

Drug synergies are ubiquitous between drugs and between drugs and other factors. Warfarin, a widely used anticoagulant preventing blood clots, interacts with everything from antidepressants to aspirin. Dietary supplements such as Saint John's wort cause reactions when paired with cigarette smoking and even cause an increased risk of bleeding when garlic is consumed. Monoamine oxidase inhibitors, used to treat depression, interact adversely and sometimes fatally with chocolate, cheese, or avocados by inducing extremely high blood pressure, while women who use birth control pills can become pregnant if they also take certain antibiotics that neutralize the pills' effectiveness. Patients who are taking cholesterol-lowering drugs such as Lipitor and who eat grapefruit put themselves at risk of liver damage or kidney failure.

The closest human example to pesticide effects on bees might be cytochrome P450, an important enzyme in both bees and humans that combats viruses, bacteria, and other foreign substances. In humans it also metabolizes drugs. Its effectiveness can be inhibited or stimulated by many pharmaceuticals, resulting in adverse or fatal effects reminiscent of how pesticide interactions affect bees.

Why do we know so much about pharmaceutical synergies? Regulatory authorities require exhaustive tests of drug interactions, and even when drugs are approved, their side effects are monitored through extensive databases reporting adverse reactions. Pharmacists are trained to inform patients of potential dangers with their prescriptions, and pill bottles are peppered with warnings.

We don't know much about pesticide synergies because the research and monitoring haven't been done. But as we've seen from the examples of recent research on bees, we need to take more seriously the possibility that we, too, are susceptible to illness or fatality from interactions between pesticides that alone are considered safe.

o   o   o

Bees may be the canary in the agricultural mine, warning us that our food systems are teetering toward collapse, victim of the human hubris that leads us to believe we are capable of managing every aspect of the world around us. Layers of management have been added over the last few hundred years, accelerating in the last one hundred years or so, so that the natural services nature provides, as resilient as they are, can no longer rebound. We need to recalibrate before other agricultural systems go the way of the bees.

Field to field, farm to farm, crop to crop, virtually every aspect of food production is on the verge of similar collapse. The ascending array of inputs necessary to manage the consequences of previous management has grown into the proverbial stack of cards ready to fall, the perfect breeding ground for synergistic collapse.

Take corn, for example, the number one legally farmed crop in the United States, exceeded economically only by illicitly grown marijuana. Corn growers apply numerous insecticides to combat a plethora of corn-munching insects, including European corn borer, corn earworm, flea beetles, fall armyworm, cutworms, and corn rootworm. For fungi, growers apply fungicides against seed rot, common and southern corn rust, northern and southern corn leaf blight, northern and gray leaf spot, anthracnose stalk rot and leaf blight, and Diplodia stalk rot, among others. Resistance by insects and fungi is epidemic, requiring frequent shifting to new pesticides as farmers attempt to keep a step ahead of their pests.

Then there are the new herbicide-tolerant, genetically modified corn varieties that allow growers to spray weedkillers while the crop is in the field. These sprays, applied at more than twelve times the volume they were ten years ago, leave fields as clean as a whistle, with nary a flowering weed to be found.

In addition, nutrition in corn is a huge issue, as it is for bees. Soil depletion over many decades of overfarming has forced farmers in the United States to use more and more fertilizer each year to replenish the soil, with the tonnage of fertilizer used annually increasing from 6.8 million tons in 1980 to 12.4 million tons in 2010, at a rate of about one thousand pounds of fertilizer required per acre, up from three hundred pounds per acre in 1960.

Finally, there's climate change and water. Increasingly unpredictable annual weather patterns in the midwestern United States require astute prediction of planting times and appropriate varieties by farmers, whose livelihood is already dependent on good guesswork. And then there's water for irrigation: the critically important Midwest's Ogallala Aquifer is being drained due to overdraws on this critical resource and, if current trends continue, will drop a further 50 percent by 2060.

Taking these factors together, the condition of corn crops appears to be remarkably similar to the condition of honeybees just before they collapsed. An increasing array of pests and diseases is leading farmers even further down the toxic chemical treadmill, inducing resistance in pest after pest. Reduced biodiversity in and around fields is a consequence of too-excellent weed control, eliminating habitat for beneficial natural pests and predators and deepening the need for pesticide applications that substitute for nature's control agents.

Pesticide residues build up in the ground and water, having further impacts individually and perhaps synergistically on beneficial organisms and perhaps even on corn itself. Depleted soil conditions force farmers to add expensive artificial fertilizers with a nutrient profile that is unbalanced in comparison to that of healthy soil. Increasing costs for these inputs require farmers to push their corn crop to even higher production to generate enough income to pay for the additives.

It's the perfect synergy storm for corn, and in fact, corn had a near-death experience in 1971, when southern leaf blight went synergistic. It's normally a minor disease, affecting less than 1 percent of the US corn crop. But in 1971 farmers planted only a few varieties of carefully selected hybrid corn, more than 80 percent of which carried a single gene that made the corn plants highly susceptible to southern leaf blight.

The interaction between blight and the susceptible seed resulted in a 15 percent crop loss that year. A permanent collapse was averted by farmers planting more diversified, nonsusceptible seeds the following year, but the recent advent of genetically modified crops has once again reduced corn biodiversity as farmers concentrate their plantings on only a few of the most successfully marketed varieties.

Most of our agricultural systems are similarly close to going over the edge. Bees have led the way, but I doubt there is a conventionally managed crop that is not on the verge of a similar synergistic breakdown.

o o o

Colony collapse disorder in honeybees was probably inevitable, a disaster for bees and beekeepers alike and for our food supply, one brought on by our best human intentions to manage the world around us. The heartbreak is personal; documentaries and news reports about CCD routinely show normally hard-nosed, practical beekeepers breaking down and sobbing. I've spent a lot of time with beekeepers, and they rarely cry in private, let alone on national television.

These normally stoic farmers are mourning their vanished bees, to which they have a deep personal attachment. But they also are lamenting the disappearance of their way of life, whether they manage tens of thousands of colonies or a modest local apiary. Small hobby and sideline operations where each colony is given tender loving care are being brought down alongside industrial beekeeping.

Solutions won't be found easily and won't be found in a continuation of megabeekeeping in a highly standardized mass agricultural system. It also won't be just hobby beekeepers in overalls tending only two or three hives and producing a few jars of honey each year. Beekeeping's future can be found in the middle ground of smaller, local operations with a few hundred hives each, integrated into an agricultural system that is compatible with stable, stationary apiaries.

We also need more diversified cropping systems interspersed with hedgerows and uncultivated areas with blooming weeds that provide a wider array of nectars and pollens than a single crop ever can. Pesticides used outside of hives will need to be tested more rigorously and used more sparingly. Beekeepers as well must use fewer chemical tools inside the hive to manage pests and diseases, or we are in for continued challenges in returning to sustainable apiaries.

Our very human tendency to simplify and seek one answer may explain our ongoing difficulty in recognizing impending synergy and acting before systems collapse. We are prone to accept death by a thousand little cuts, in which one degraded aspect of our environment or health becomes familiar and accepted as normal—and then another.

Sooner or later, though, there is that thousandth cut, insignificant on its own but deadly in the context of many other cuts. That's what's happened to bees. Myriad afflictions, each controllable alone by bees' natural resilience, finally crossed the line where synergy kicks in.

The ensuing interactions turned a few small cuts into a hemorrhage, a tragedy but also an opportunity to consider what we might learn from the collapse of the bees. Complex systems like bee colonies or human societies are marvelously functional when healthy, but their very complexity makes social organisms particularly vulnerable to sudden and epic breakdowns when overcome by too many challenges.

Yves LeConte and I exchanged e-mails a few months ago, and he had this to say about the future of bees and beekeeping: "I think that the bees should really be supported at the moment as the level of colony mortalities is way too high. But we need pesticides, so what to do? In France, most people here want to protect the bees, which is now a kind of symbol of our environment to be protected. It is now in the hands of politicians . . . My feeling is also that most of the bee colonies are in the hands of the beekeepers, so we need to support them so that they do not disappear [along with] the bees they manage."

Our human ingenuity is remarkably effective at solving individual problems, but our blind spot is our inability to recognize when interactions between issues might bring us down.

We failed to see it coming with our bees, with catastrophic consequences for these marvelous insects and the systems that depend on them.

But the deepest tragedy may be this: beekeepers and civilians alike connect with bees at that deep, unfathomable place where nature moves our emotions. As bees collapse, so does our opportunity to appreciate a creature that, like us, has survived and prospered through collaboration and relationships with others.

Their demise, and the synergy behind it, does not bode well for our human future.

5

Valuing Nature

Extending thousands of miles from Labrador in the east to Alaska in the west, Canada's northernmost boreal forest is among the most spectacular ecosystems on Earth. It is the largest intact forest on our planet, rich in black spruce, white birch, balsam fir, jack pine, and trembling aspen, home to woodland caribou, moose, beaver, and lake trout. Interspersed with the forest are extensive wetlands that in summer house a good portion of the world's migratory birds.

This region of long winters and short, mild summers is also rich in minerals, oil, gas, lumber, and hydroelectric power. Extraction of these forest riches has created much wealth but at the same time has left extensive environmental damage. Today, extracting oil from the bitumen-rich tar sands has become the largest mining project on the planet, generating considerable controversy as the impact of mining this hard-to-extract resource widens across the forests of Alberta and

Saskatchewan and spreads globally via the increased greenhouse gas production associated with removing bitumen and processing it into oil.

The southern edge of the boreal forest is also the northern limit of farming in North America. Given the short growing season, poor soil, and ferocious biting flies and mosquitoes that plague inhabitants, it's remarkable that crops can be farmed at all. It was here on farming's geographic fringe that, beginning in 2002, my laboratory conducted one of the first ecosystem-based studies of the economic impact of wild bees on agriculture.

Our focus was on discerning the effects of various crop-raising systems on wild bees and, conversely, whether wild bees could be sufficiently diverse and abundant to pollinate crops. We wanted to find out whether unmanaged habitats could provide enough service from beneficial wild pollinators to supplement or even displace the managed honeybees that have largely replaced them in agroecosystems.

My student Lora Morandin studied wild bees foraging on canola, a crop that is itself an example of human ingenuity. Canola is a product of science, created by Agriculture Canada scientists in the early 1970s by breeding undesirable qualities out of its predecessor, rape. They bred an oilseed crop that yielded edible oil rather than the machine oils extracted from rapeseed.

Canola requires bee pollination, but few wild bees remain in the heavily farmed areas throughout most of the crop's range. Heavy insecticide use, destruction of nesting sites as fields are tilled, and extensive single-crop plantings that eliminate forage for bees outside the crop have taken a serious toll on wild bees not only on canola but also in most farming-intensive habitats. As a result, growers have to rent managed honeybee colonies that are moved in each year when the crop blooms and moved out after flowering.

Lora worked in northern Alberta, at the southern edge of the boreal woods, a landscape mosaic of farmed fields, aspen woodland, grassland, shrubland, wetland, and cattle pasture adjacent to the Peace River. We chose this area hoping it would be a best-case habitat, a mix of highly disturbed and unmanaged land near farms that might still support abundant populations of wild bees. It is one of the few areas in North America where a small number of canola farms qualify as organic since they are isolated enough to be unaffected by the insecticide use and genetically modified (GM) crops that are abundant in most agricultural communities. And untypically, growers didn't yet rent honeybees but instead relied on wild bees to pollinate their canola.

Lora compared three types of canola fields: organic, conventional, and genetically modified. The organic fields adhered strictly to regulatory guidelines that prohibit most pesticides. Conventional fields were sprayed with insecticides, fungicides, and/or herbicides. The GM canola was herbicide resistant, meaning that weedkiller could be sprayed in the fields while canola was present, a system that is more effective at killing weeds than conventional farming, in which herbicides are sprayed before planting or after harvest.

She found the highest numbers of wild bees and the best pollination in the organic fields, with conventional farms intermediate and the GM canola the lowest in bee numbers and the effectiveness of their pollination. Further, bee abundance and diversity were clearly connected to the abundance and diversity of weeds in and near the fields and—on a more landscape-level scale—to the amount of unmanaged habitat adjacent to or within bee flight of the crop.

Most remarkably, Lora did an economic analysis that showed yield and profit could be maximized if about 30 percent of land was left uncultivated. Farmers who planted 100 percent of their fields would earn about $27,000 in profit per farm, whereas those who left 33 percent unmanaged would earn around $65,000 in profits, all due to improved pollination and the resultant increase in yield. These were startling, counterintuitive results: up to a point, planting less acreage resulted in higher yields and greater income simply by relying on unmanaged land to increase wild bee populations.

This study was prescient in addressing what has become a critical issue in world agriculture: how will growers pollinate crops as managed honeybees decline in numbers and increase in cost?

Lora's study may have been conducted at the geographic fringe of farming, but it addressed the most central of questions: can habitats in and near farmland be restored to support sufficient populations of wild bees to be economically viable pollinators?

o o o

Thousands of miles south of our Alberta research site sits California's Central Valley, possibly the most extensive manicured habitat on Earth, where much of the world's food is grown, including more than four hundred crops that require or benefit from bee pollination. Stretching 450 miles through central California, this flat valley between the Coast and the Sierra Nevada mountain ranges provides the illusion that nature can be tamed and organized for human benefit into a permanently green garden, without consequences for the natural world.

Verlyn Klinkenborg, a well-known writer on agricultural issues and member of the *New York Times* editorial board, wrote about the Central Valley's "endless rigid, rectified miles of trees and vines and seed rows. There is something stunning in the way the soil has been engineered into precision. Every human imperfection linked with the word 'farming' has been erased. The rows are machined. The earth is molded . . . This is no longer soil. It is infrastructure, a biological desert, a place where only a handful of species are allowed to thrive."

The valley's profuse productivity and highly arranged farmland leave the impression of good farming, but that depends on what you think good farming means. That tableau is a consequence of massive chemical inputs of pesticides and fertilizer, a flood of irrigation water that is draining aquifers and surface water from the western half of the continent, and huge acreages of single-crop plantings with significant impact on biodiversity.

I spent a few days there in June of 2013 visiting bee researchers and almond growers, traveling daily between the huge megalopolis of the Bay Area and the equally engineered megafarms of the valley. It was a disjointing pendulum of food experiences. In the Bay Area I had many home-cooked and restaurant meals with friends whose food ethic was focused on local, organic, vegan, slow food, and artisanal cooking. Then I would go to the valley, where farming is predominately mechanized and industrial, out of touch with the food movements that are sweeping urban regions.

I traveled to California's Central Valley because almonds have been particularly hard hit by shortages of honeybees, and 80 percent of the world's almond production takes place here. It's the second-largest valley crop after dairy and the state's number one export, forecast at $3.8 billion in 2013.

Almonds, a close relative of peaches, are one of the world's oldest crops, cultivated across its native Middle Eastern to South Asian range as early as 3000 BCE. It's one of those crops where it's hard to fathom how anyone got the idea to eat it; almonds in their wild form are bitter from cyanide, at concentrations high enough that a few dozen wild almonds are deadly. But somehow early growers selected from the wild stock to produce a sweeter form without cyanide, possibly due to a genetic mutation, and the almond industry was born. Almonds were brought to California in the mid-1700s by Franciscan padres and today occupy eight hundred thousand acres in the Central Valley. The nuts are sold raw or toasted, plain or flavored with anything from hickory smoke to tamari sauce, and are also converted into almond milk, flour, butter, paste, and oil.

The trees bloom in February, their flowers producing pollen that attracts bees, but not much nectar. Almonds are both a high-value crop and one that requires bee pollination to set seed. Growers are willing to pay a premium price to bring 60 percent of the honeybee colonies in the United States to California when the almonds flower.

That's 1.5 million colonies in 2013, trucked in from as far away as Florida at a cost to growers of about $300 million. Stress from this late-winter mass migration is a major contributor to colony collapse disorder. Almond growers are as worried as beekeepers. Each year for almost the last decade, it's been an increasing challenge to get enough bees to pollinate their crops. The colony shortages have reached crisis proportions, serious enough to catch the attention of major national media, including a thirty-five-minute television feature on *Dan Rather Reports*.

o o o

Almonds are like most orchard crops in requiring patience, a trait found in many of the multigenerational farm families that still make up a considerable portion of the Central Valley's producers. I visited Drew Scofield and his son, Tyler, who represent the third and fourth generations on their family-run farm.

Drew didn't want to become a farmer. His parents told him it was a tough life and to go do something else. He played basketball at Sacramento State College, but just when he graduated, he got a call from one of his high school classmates: "She said, 'Well, my husband left me, and I need some help on the farm.' I said, 'Well, I'll show up on Monday, and I'll help you however I can.' So I showed up, and she said 'OK, here's the farm. You farm it and give us a percentage of the crop,' and that's really how I got started." He took over his family's farm a few years later.

Drew is tall and wiry, brown, and wrinkled from a lifetime in the sun, his back "shot to hell" from way too much farm work. Tyler is a shorter, younger version of Drew, not quite as dark as his dad, with more hair and a still-intact back. We talked in their farmhouse, decorated with rusty antique farm implements on the walls. Although the house was neat and clean, it was clear that they pay more attention to the farm.

Claire Brittain, a bee researcher from the University of California at Davis, had arranged our visit. Drew and Tyler greeted us most hospitably with fresh-squeezed juice made from oranges picked minutes before in their hobby citrus orchard just outside their front door. When we left, they gave us boxes of fruit, exotic cucumbers, and blueberries to take home.

I've met a lot of farmers, and though they may be highly variable in how they farm and many aspects of their personalities, all the good ones exhibit a suite of related characteristics suited to farm life. The best farmers are persistent, practical, resilient, flexible, and committed to doing whatever it takes to get the job done.

Drew and Tyler are no exception, and as a father-son team they habitually start and finish each other's sentences. They pronounce almonds as "ay-mends," rhyming with "amen," and note that it's a crop requiring a particularly long view. It takes up to six years for a new orchard to begin yielding and twenty to twenty-five years before the orchard needs replanting.

I asked them why almonds did so well in the Central Valley, and together they replied: "Well, it was a natural here, in this particular area especially. In February it's warm right here; the climate, soil, the irrigation that was put in and the bottom line, the dollar, it all adds up." Generations of growing almonds have taught them a lot about farming: "You gotta have patience. It's not an instant reward. It's a cycle. You pay the dues initially. It's a cycle of taking out the old and planting the new . . ." It's essential, they say, to "watch the bottom line more than ever—things can slip away. Do it right the first time, and be prepared to do what it takes to get it right."

The Scofields were unusually prophetic about the current honeybee shortages. Decades ago, beekeepers would ask growers to let them leave their bees near almonds over the winter, free of charge, but increased tree densities and acreages shifted that relationship to one where growers needed the bee colonies. Beekeepers started charging growers rental fees in the 1970s.

The Scofields saw problems developing with bee supply as far back as ten years ago, as beekeepers would fail at the last minute to appear with colonies: "I'd get a call from a yahoo beekeeper: 'Hey, your bees have died.' I had enough of that." Drew and Tyler didn't wait around. With their characteristic get-'er-done attitude, they bought their own honeybees and now manage eight hundred colonies not for honey production but purely to pollinate their almonds and other crops. It's a risk that has paid off financially; the price of a honeybee colony has tripled since 2003, to up to $200 for even mediocre hives in 2013, and makes up between 16 and 25 percent of growers' annual costs.

Drew and Tyler also did something else unusual: they planted hedgerows, creekbeds, and irrigation ditches with bee-friendly plants. Their noncropland was easily the richest unfarmed habitat I saw in my two days in the valley; when I visited, three or four varieties of lavender were in full bloom, the flowers laden with bees collecting nectar.

Farmers habitually consider noncrop plants to be undesirable weeds, competing for water, nutrients, and pollinators with the crops. The typical almond grower smites weeds with herbicides, but the Scofields have taken a different approach: "We plant creeksides, riparian areas, get something blooming every month of the year. Acreage-wise, it's not much, but I don't know of anyone who plants native or beneficial plants like we do. We have a lot of acreage we can't farm [but] still pay taxes on, so we try to do something."

Most other almond growers became concerned about the impending honeybee crisis later than the Scofields, but instead of keeping honeybees they are pinning their hopes on plant breeders experimenting with self-pollinating varieties of almonds that do not require bees. The US Department of Agriculture has been working with a variety called Tuono from Spain, but although it will set seed without bees, it has a hairy seed coat, thicker shell, and half the edible nut of bee-pollinated varieties.

A private nursery in California bred another variety, Independence. Some growers are beginning to plant acreages of Independence as a buffer against bee shortages. But it, too, produces less edible nut and has a taste that's different from pollinated varieties. It remains unclear whether scientists and horticulturists can breed a commercially successful selfing variety. Still, even a 10 percent shift in acreage to self-pollinating varieties from those that require bees would help reduce the high demand for honeybee colonies.

Honeybee shortages are getting extreme enough that some growers are considering desperate measures, including using tiny insect-size microaerial vehicles with flapping wings originally developed by the military for covert sensing. It's been proposed, seriously, to send fleets of them into the orchards to seek flowers in bloom, land, and then move between flowers and deposit pollen like bees.

But there's another idea that's not so wacky, one that has gained considerable traction in the research community and is beginning to attract the attention of progressive growers: enhancing populations of wild bees. It's a key idea that may determine whether almonds remain a top crop in the Central Valley: can nature's services be harnessed to supplement or replace managed honeybees with wild bees as pollinators?

The issue here is larger than just bees, as important as that is. At stake are two competing visions for how we grow food: the current paradigm, which relies on industrial scales and extreme inputs, and a more earth-friendly variety that maintains ecosystem services and promotes natural and crop biodiversity.

What's in question here is the idea of sustainability and whether it's just a buzzword or a concept that can be made real in agriculture.

o o o

There are about fifteen hundred wild bee species in California, a feral community that could provide a basis for ecosystem-based pollination rather than renting managed bees. The bees are diverse, ranging in size from tiny to among the largest flying insects in the world, some with very short and others with quite long tongues that are well adapted for the length of particular flowers. Some emerge from hibernation to build nests before spring, while others are active into the fall.

Their diversity is stunning, at least to entomologists who take the time to look closely at their anatomical features and lifestyles. This variety is important because it means that there are wild bees that potentially match well with each crop, which in turn suggests that the raw material is there in nature to enhance wild bee populations and diminish the pollination crisis.

Whatever their size, shape, or habits, all bee-pollinated plants are subject to basically similar pollination. The bee visits a flower, picks up pollen on its hairy body, and incidentally deposits it on the next flower it forages on. The pollen (sperm) makes it way to fertilize the plant's ovule (egg), and seed is set.

Bees can improve crop yields in a number of ways, depending on the particular characteristics of the plants. Some are pollinated by one bee visit, while those with multiple seeds take many. Some crops absolutely require bee visitation, whereas others benefit but can also self-pollinate or set seed from wind pollination. Honeybees are decent pollinators for many crops, but for most crops one or more wild bee species are more effective on a bee-by-bee basis.

The first step in addressing whether wild bees could be effective crop pollinators was to ask the most fundamental questions: What kinds of habitats enhance wild pollinators? Do wild bees visit and pollinate early-blooming almonds and other crops? If so, what factors would increase their diversity and abundance to provide a commercially viable service?

Two research teams from two quite different California universities have devoted the last few years to answering these deceptively simple questions. One, led by Claire Kremen, is at Berkeley, an institution still flavored by the political upheavals of the 1960s and 1970s. The second, led by Neal Williams at Davis, is smack in the middle of the Central Valley, with deep ties to the adjacent farming communities, and is very much considered an aggie school.

The two research groups have subtle differences in how they approach their work but in many ways are as hard to differentiate as closely related wild bees are to a nonexpert.

Kremen's and Williams's laboratories meet frequently, coauthor publications regularly, share resources, and often swap students and research assistants. Collectively they have assembled the most comprehensive dataset for any global agroecosystem demonstrating the potential for and the limitations of wild pollinators to supplement or replace managed honeybees.

Their results have echoed those of our work on canola in northern Alberta, indicating that greater wild pollinator diversity and abundance are associated with enhanced pollination, better fruit set, and higher yields in almonds. More extensive areas of natural habitat near crops are associated with greater pollinator diversity. Wild bees are almost absent in orchards without adjacent natural or seminatural habitats.

They've also investigated the types of habitats that best enhance free-living pollinator populations. Natural, unmanaged habitats close to orchards are finest, but managed plantings of vegetation attractive to wild bees can also increase pollinator numbers and efficacy.

Hedgerows, low shrubs, bushes, and underlying vegetation that use little land and can fit into edges or marginal areas on farms are the most effective plantings. Native flowering plants are preferred since they best support the native bees, with which they coevolved. Effective bee plants include lilacs, buckwheat, wild rose, coffeeberry, coyote brush, Mexican elderberry, toyon, sage, poppy, gumplant, yarrow, and quail bush. Strips of lower-lying weedy plants also increase pollinators, but pollinators near these rows tend to forage more on the edges of orchards and don't move as deeply into the interior, possibly because the weeds distract them from the orchards.

Another research question has been whether honeybees or wild bees are better pollinators. The results have been pleasing to those of us who don't want to take sides in the "which bee is best" discussion. It turns out that honeybees and wild pollinators each have subtly different foraging tendencies that complement each other rather than compete.

Honeybees tend to forage higher on the tree, and wild bees are evenly dispersed throughout tree canopies. Wild bees fly during stronger winds, colder temperatures, and more inclement weather than honeybees, even in light rain, making them a potentially significant contributor to pollination for almonds, which bloom early in the season, when high winds, cold temperatures, and rain are common. Perhaps most remarkably, the presence of wild bees actually enhances the effectiveness of managed honeybees.

Almond trees are planted in rows that alternate between varieties since fruit set and yield are improved when almond flowers are pollinated by pollen from a different variety. This system benefits from bees moving between rows to effect the best pollination, but honeybees tend to move compulsively down single rows rather than cross the space between rows.

Wild bees, on the other hand, move laterally between rows more readily than honeybees, but the presence of wild bees also induces honeybees to travel more between rows, improving their effectiveness as pollinators. Similar results have been found in sunflowers, although why wild bees inspire honeybees to forage laterally rather than straight along almond or sunflower rows and why honeybees tend to forage in straight rows have yet to be determined.

A similarly positive picture of the potential for wild bees to pollinate diverse crops is emerging globally, confirming the Central Valley research indicating that unmanaged vegetation or bee-friendly noncrop plantings can support sufficient wild pollinators to supplement or even substitute for honeybees. One massive forty-nine-author study was published in the prestigious journal *Science* in 2013. It synthesized data from forty-one crop systems worldwide, collected from six hundred fields on all continents except Antarctica, and included fruit, seed, and nut crops that benefit from or depend on insect pollination. The authors did a meta-analysis, a set of statistical methods that reveal patterns from many studies, commonly used to analyze multiple clinical trials of a new drug or medical treatment.

They found increased fruit set with wild-insect visits on all forty-one crop systems, but fruit set increased significantly with honey bee visits in only 14 percent of the systems surveyed, indicating that, when present, wild bees pollinate crops more effectively. Their results "suggest that new practices for integrated management of both honey bees and diverse wild-insect assemblages will enhance global crop yields." The authors propose turning pollination management on its head and consider honeybees rather than wild bees to be the supplemental pollinators. Nevertheless, they recognize that honeybees will likely remain as partners in pollinating most crops as they can be provided in larger numbers than wild bees, trading off quantity of bees against the often-superior quality of wild bees individually.

Management to enhance wild pollinators needs to be specifically tailored to each crop but generally would include conservation or restoration of natural or seminatural areas within and near croplands, addition of diverse floral and nesting resources, and reduction of pesticide impact on wild pollinators. The take-home story from this growing body of research is that wild bees will not necessarily replace honeybees, but they can certainly reduce our dependence on managed pollinators. Both enhancing the area of unmanaged vegetation and planting bee-friendly hedgerows and/or ground cover on and around farms increase the diversity and abundance of free-living pollinators, indicating that creating a more diversified pollination system is feasible.

But the practices of modern industrial farming are not consistent with strategies to enhance wild pollinators. Vast monocropped acreages that utilize every square inch of available area, intense use of insecticides that often are toxic to bees, massive herbicide sprays that remove vital weed forage from fields and orchards, and intensive tilling, which disrupts nesting sites, are harsh practices from a bee's point of view.

And pollination is only one factor in how we grow food. Disagreements about how to farm tend to polarize into organic versus conventional, but that may not be the most meaningful divide. Rather, the key issues may revolve more around diversity versus monoculture and whether we create farms that rely on ecosystem services and mixed cropping or continue to depend on large-scale, single-crop farming with massive, expensive inputs of chemicals, fertilizer, and managed honeybees.

Many view the honeybee crisis as the canary in the mine for contemporary agriculture, an early warning sign that business as usual on the farm may be reaching a crisis point. Claire Kremen and Neal Williams see it that way. They have taken their insights into pollination and moved deeper, developing a view of a diversified pollination system that may be the harbinger of a much-needed revolution in how we grow food.

o   o   o

Kremen's academic home, the University of California at Berkeley, looks superficially like a picture-perfect university campus. Shaded walkways with towering old trees, expanses of green grass cluttered with clusters of earnest students in animated conversation, and imposing formal stone buildings are mixed in with modern glass construction.

A closer look reveals that severe government austerity has taken its toll; the physical campus has seen better days.

Kremen's large office is in a repurposed older stone building, Wellman Hall, a landmark building originally designed for agricultural research. Its high ceilings and solid wood furniture reflect the hall's classic construction, but the ceilings are peeling, and the furniture is worn, stained, and in desperate need of refinishing.

Kremen works in the College of Natural Resources, itself repurposed from its original agricultural orientation to one that, its website notes, "addresses biological, social, and economic challenges associated with protecting natural resources and the environment." It's a perspective more focused on ecological sustainability than growing food and serves as a useful counterbalance to the traditional agricultural universities, which tend to focus on maximizing production.

Kremen herself is focused and articulate and speaks in complete sentences that connect the dots between related ideas. She grew up in North Carolina and attended Stanford and then Duke. At Duke she had her career epiphany: what she wanted to do with her life was reduce world hunger. She was also attracted to biodiversity and conservation issues. For a while she turned her attention from world hunger to studying butterflies in Madagascar. But when she first heard the term "ecosystem services," she realized it was the missing link between her concern about hunger and her passion for conserving biodiversity.

We met one June morning in her office, its shelves stacked with books on conservation, evolution, and bees, walls decorated with "Save the Pollinator" posters, her whiteboard covered with recently drawn diagrams of connected boxes containing phrases that perfectly summarize her lab's interests: Economic Outreach, Land Management, Ecosystem Function, Yield. After a few hours we switched venues to her laboratory's field vehicle and drove around the Central Valley looking at research sites an hour or two outside of Berkeley.

The takeoff point for Kremen's work is wild bees and determining whether feral populations can be enhanced sufficiently to pollinate crops effectively in the Central Valley, an agroecosystem that is harsh for pollinators: "The way that we produce food, the dominant model for food production, is this industrialized model that is really bad for wild bees. The system produces bloom for only a couple of weeks a year. During the rest of the year it's pretty barren, no crop blooming, no weeds blooming. Monocultures are really not very good for bees."

Results from her hedgerow research clearly demonstrate that wild pollinators can contribute, but the question is how much and also whether farmers are willing to implement the habitat changes necessary to accommodate wild bees. Kremen is realistic about the potential: "[The use of hedgerows] definitely improves pollinators, but it's not sufficient for the pollination services we need in crops."

It's largely a problem of scale: "Right now agriculture in California is dominated by huge, huge swathes of very, very large fields generally growing one thing on a very, very large scale. We're looking at a very, very tiny adjustment to that landscape, placing a strip of native vegetation along one edge of the field. Even if we were to place hedgerows around every single edge of every field, it's still a tiny change that's probably not going to be enough in the long run. Hedgerow work is futile in and of itself. It's not a silver bullet. It's only one piece of what needs to be done. I don't believe it's enough."

It's unusual for a researcher to diminish rather than exaggerate the significance of her work, but for Kremen the hedgerow research is just part of a much more comprehensive vision for a well-pollinated Central Valley: "I think we need a radical transformation of our food system, one that is ecosystem service based. We need a much more integrated system where we have diversity established at many different levels, many different spatial and temporal scales—a whole landscape that's diversified, in which each of the farms in it is also diversified, a polyculture within each farm, with patches of natural habitat integrated to provide habitat for beneficial insects."

Neal Williams also sees the need for diverse approaches, and from his faculty position at an old-style agricultural university, University of California–Davis, he understands the barriers that need to be overcome before ecosystem services make significant inroads in reducing the inputs that fuel industrial farming.

We talked at the Harry H. Laidlaw Jr. Honey Bee Research Facility, a short drive from the Davis main campus. It was named after a UC–Davis professor renowned for his landmark research in honeybee genetics, exceptional reputation for generosity, and notably gentlemanly demeanor. I knew the facility well; I had slept outside the complex in a hammock strung between two trees as a graduate student during my first visit in 1975 to learn more about honeybees before embarking for French Guiana to study African bees.

With basic training in evolution and ecology rather than the more applied foot-in-furrow mindset that still dominates research at agricultural universities, Williams represents a new generation of faculty at Davis. His training is in basic aspects of insect foraging behavior, but he chose to work on wild bees because he wanted to do something of public importance, and marrying fundamental elements of bee foraging behavior with practical applications seemed a natural fit.

He and Kremen have reached similar conclusions about bees and the broader need to diversify habitats in and around cropland. They both propose four components that would move farming toward relying more on ecosystem services: leaving some land in a natural, unmanaged state; planting hedgerows along edges and irrigation ditches; providing strips of low-growing, bee-attractive weeds in and around crops; and diversifying farming into more mixed-crop polycultures rather than single-crop monocultures. For almonds, they suggest focusing on plantings that bloom just before and just after almonds.

Williams also talked with me about farmers, who exhibit an idiosyncratic mix of high tolerance for risk and conservative attitudes toward change. As a group farmers have been interested in and supportive of wild bee research but are not moving quickly toward the obvious solution to the bee crisis: diversifying pollinators.

Planting weeds and taking up valuable cropland with non-crop plantings are different from what farmers are used to doing. Williams told me that "One of the things we always struggle with in these more applied questions is that we would like to move the community toward embracing something different from what they are doing, a more sustainable, diversified agricultural landscape, and there's resistance. There also seems to be this strong sense that uncultivated land is not being wisely used. The only land really available are these ribbons of irrigation ditches that run through the Central Valley, or road verges, but the problem there is that they're all thin lines barely noticeable in an immense farmed area. There's resistance because that's not what they do: it's an irrigation canal, so of course we don't want anything growing on it. You're asking them to do something you're not going to pay them to do."

Farmers are also concerned that hedgerows and unmanaged vegetation will take up water, a commodity already in scarce supply in the Central Valley, and may harbor birds and rodents that might eat seeds. Further, producers and everyone up the chain of food distribution are terrified of litigation and worry that birds and rodents could defecate in the fields and spread salmonella or *E. coli* bacteria.

There's no evidence this is the case, and farms have coexisted with nature for millennia without spreading disease. Still, pressure from the large chain groceries has made farmers want to reduce even unproven risks, and recent spinach contamination scares with *E. coli* have added to the sensitivity, although there's no reason to blame ecosystem services. Most experts attribute farm-based spread of diseases to lack of portable toilets and hand-washing stations for field workers, proximity to poultry farms, and the use of chemical fertilizers, compost, and manure.

One option, of course, is to let the marketplace decide what steps, if any, farmers should take in response to the diminishing capacity of managed honeybees to pollinate their crops. But there's more in play here than just the livelihood of farmers or beekeepers.

There's a food revolution going on all across North America, with its call-to-action phrases like "local," "farm-to-fork," "artisanal," and its lead mantra, "sustainable." This compelling vision of ecosystem services is integrated with a more human scale of food production.

The only problem is that most farmers haven't yet bought into the message, especially in the Central Valley of California.

o   o   o

Perhaps the most overworked term in the lexicon of public discourse today is "sustainability." It's a word vested with hope and a noble goal: humans should live lightly and in harmony with the natural world around us and with each other. But it's a vague word to apply to ecological, economic, social, and cultural parameters that touch on virtually every aspect of human life.

Whatever sustainability is, we haven't yet achieved it, and that is acutely the case for the vast majority of North American agriculture. Advocacy from progressive urban foodies has not bubbled up to most farms, which remain productive because of massive inputs, industrial-scale management, and tolerance for severe environmental impacts.

There's been progress; major supermarkets have at least a small organic section; a few national grocery chains such as Whole Foods focus on organic and sustainable products; and there's been explosive growth in the number of farmers' markets in most North American cities over the last five to ten years, through which urbanites can buy farm-fresh food from regional growers. But we have not yet reached that tipping point where these small changes and innovations have accelerated into large-scale agricultural change. In this way farming is similar to other progressive innovations such as hybrid cars, wind and solar energy, and urban composting: they're all growing slowly, but none have yet become dominant practices.

Experiments with alternative approaches to pollination provide a glimpse into what sustainable farming might look like and through that vista reveal some beachhead initiatives that may begin to leverage us into that holy grail of sustainable agriculture.

Research is one area where positive results can go viral and tip us more quickly toward sustainable options. Studies on wild bees have traditionally been the seriously underfunded cousin of honeybee research, a microcosm of the broader disparity between the vast government-supported empire of conventional farming and an ecosystem-based alternative. Claire Kremen, for example, has never received funding from the most obvious source, the US Department of Agriculture. Oddly, most of her funds come from the bulging coffers of the Pentagon, in grants from the Army Research Office, which is justified in protecting food security as a defense issue rather than through any military interest in sustainability.

This is changing, and one large project based at Michigan State University provides a model for a large-scale, collaborative, national, and well-funded initiative that can plant seeds of change throughout the agroecosystem. The Integrated Crop Pollination (ICP) project is a big scale-up for the wild pollinator community. It's half-funded by the USDA, and various public and private sources provide the other half, with a total budget of $18.8 million over five years. Sixteen research labs across the country are involved, including dozens of students, faculty, postdoctoral fellows, research staff, and commodity groups.

The mandate of the ICP is to investigate "the combined use of different pollinator species, habitat augmentation, and crop management practices to provide reliable and economical pollination of crops." The consortium is in the first year of studying the performance, economics, and farmer perceptions of different pollination strategies in several fruit and vegetable crops and—unusually for a research project—has a strong component of outreach to farmers.

But there's sufficient knowledge to begin implementing pollinator habitat-enhancement programs, particularly in California's Central Valley. The key is to integrate cropland with natural habitat. In this respect action remains far behind the research. The primary US organization involved in habitat enhancement in and around farms is the Natural Resources Conservation Service (NRCS), a branch of the USDA. It's the former Soil Conservation Service, prominent during the Dust Bowl era of the 1930s, and it still maintains more of a focus on preserving soil than conserving or restoring what grows above the ground.

They do provide information to farmers about preserving pollinators and some financial benefits to farmers who create or enhance pollinator habitat. But the NRCS is not as effective as it could be because it only partially funds habitat renovation and, like many government programs, requires complex application forms and background information beyond the time and resources available to most farmers. Organic farmers have been the most common applicants, disappointing since the program ends up supporting the already converted rather than easing new growers into the ecosystem services fold.

Perusal of the total USDA subsidy budget over the last few decades reveals a disturbing pattern of propping up conventional agriculture, which needs it the least, rather than support for ecosystem services, which represent a more holistic approach to farming. From 1995 to 2012 the USDA paid out $295 billion in subsidies, the vast majority to increase revenue for particular commodity growers, especially the four most highly subsidized crops: corn, wheat, cotton, and soybeans. Habitat renovation received a miniscule amount, only $4 billion, and most of that went to preserving wetlands.

Still, there are slight signs of change. In February 2014 the USDA announced a $3 million program to provide guidance and support to farmers interested in diversifying food sources for honeybees. Although focused on managed honeybees, wild bees will also benefit from the program.

o   o   o

What would it take for pollination to become ecologically based rather than management based? Research demonstrating that ecosystem services best meet the economic interests of farmers, consumer demand for food grown under more sustainable circumstances, and changes in government policy and funding are needed to get us to where ecosystem services might become the primary organizing model for crop pollination.

The research is already there, although it needs to be expanded to more crops. Where ecosystem services have been studied, from northern Alberta to the hot Central Valley of California and elsewhere, the results have consistently indicated clear and quantifiable economic benefits to habitat conservation and/or restoration of crop pollination.

Consumer demand for bee conservation is the second stimulus toward ecosystem services that would lead to sustainable agriculture, and it is on a steep growth curve. The urban food revolution is a prominent component of city culture in North America today: consumers are paying more attention to where our food comes from, how it is grown, and whether it comes to us as an industrial or an ecological product.

A few programs are attempting to scale up this enthusiasm into an effective movement, such as a Canadian program called Local Food Plus and the Portland, Oregon, group Food Alliance. Both are run by nonprofit societies that certify and label food as being more sustainable than the products of conventional farming but not fully organic. Hundreds of farmers, restaurants, grocery stores, food processors, caterers, and institutions have qualified for and proudly display the Local Food Plus or Food Alliance labels, but the tipping point where sustainability labeling is widespread is still far in the future.

The criteria for identifying sustainable products include having come from farms that reduce or eliminate synthetic pesticides and fertilizers, avoid hormones and antibiotics, conserve soil and water, provide safe working conditions for farm labor, reduce on-farm energy consumption and greenhouse gas emissions, and protect and enhance wildlife habitat and biodiversity on working farm landscapes. There's no "pollinated primarily by wild bees" component specifically, but many of the qualities needed for certification would certainly encourage free-living, unmanaged pollinators.

Local Food Plus and Food Alliance appear to be North America's only two sustainable food certification systems, and both are primarily private enterprises, albeit nonprofit ones.

The lack of government involvement at any level is palpable, both in setting effective policy and funding, although almost every government office associated with agriculture prominently claims to be fostering sustainability.

Approaches to conserving wild bees also tend to be strongly citizen based, particularly in Europe and the United Kingdom, rather than government inspired, exemplified by the United Kingdom's Bumblebee Conservation Trust. This nonprofit has grown to seven thousand members and a full-time staff of thirteen, with testimonial support from some of the United Kingdom's best-known celebrities, including television and print superstar David Attenborough and television gardener Toby Buckland. The trust's work focuses on raising public awareness, influencing policies that might affect wild bees, and perhaps most significantly the preservation, creation, and restoration of floral-rich habitats.

Its members envision and are working toward "a different future in which our communities and countryside are rich in bumblebees and colorful flowers, supporting a diversity of wildlife and habitats for everyone to enjoy." It's an enthusiastic group, but it has a considerable amount of habitat loss to overcome. Members estimate that Britain has lost 97 percent of its meadowlands, more than seven million acres, since 1930, and they recognize it's going to be a long slog to success.

Still, if we are ever to tip the scales toward sustainable agriculture and, more specifically, develop an effective, diverse pollinator community, government is going to have to do more. It is, after all, government policies and agricultural funding that have facilitated today's management-heavy farming system, which requires huge inputs and massive government support to survive.

And what about the growers themselves, who are the ultimate arbitrator about what grows on their farms and how it's grown? Everyone I talked with from producers like the

Scofields to research scientists said basically the same thing, best expressed by Franz Niederholzer, a California farm advisor I talked with: "It's a 'you do what you know' kind of thing."

Most farmers, unlike the Scofields, don't know bees and view them as managed tools brought onto and removed from their farms by suppliers rather than seeing bees as a biological service. A growers' instinct when bee shortages occur is to plant a variety that doesn't need pollination rather than to create a more bee-friendly habitat.

o o o

Ironically, colony collapse disorder in honeybees could be a blessing for both beekeepers and farmers. It is and will continue to be a painful transition, but the shocking demise of managed honeybees has brought a major issue in agriculture to public attention: When does our quintessentially human predilection to manage the world around us become toxic, and can the promise of a more balanced, sustainable system overcome the inertia of industrial agriculture?

The pollination crisis in almonds and elsewhere is really about management—about how we manage honeybees and how we manage the habitats on which both honeybees and free-living wild pollinators depend. We are frustratingly close to a sustainable pollination system based on the ecosystem services of wild bees, yet we are still trapped by the inertia of past practices and our human habit of wanting to fix problems through management rather than seeking a less action-based natural balance.

Because of recent research stimulated by colony collapse disorder in honeybees, it's now possible to envision just how simple and practical it would be to transition to a more sustainable mix of wild pollinators and managed honeybees. The details vary from crop to crop, but the core elements are similar.

This model pollinator service would consider feral pollinators as primary, with honeybees supplemental. Polyculture, through its carefully planned plantings of crops in smaller acreages as well as the expansion of unmanaged and planted habitats in and around cropland, would provide more diverse forage for both wild and managed bees, as well as enhanced nesting sites for free-living pollinators.

A sustainable pollinator system would also be characterized by reduced pesticide use, both the insecticides that are toxic to bees and the herbicides that kill weeds that maintain wild bees. Finally, we need to reduce the mass migration of honeybees all around North America, relying more on local apiaries in which colonies stay put and taking advantage of the abundant floral resources that would result from redesigned polyculture and enhanced habitat agroecosystems.

Ironically, successful wild pollinators might just save the honeybees. Mass movements of colonies across North America worked in the past but are no longer viable due to the added stresses of diseases, pesticides on farms as well as those applied within hives by beekeepers, and the same nutritional challenges of monocropped acreages that confront wild bees. It's just too much, and we need to regroup with a simpler, less intensive, and less mobile way of keeping honeybees.

If we can succeed in shifting crop pollination toward a more habitat-based system, perhaps that lesson will spill over into reducing the other excessive ways we manage agricultural environments. The achievable vision of a sustainable pollinator community reminds us more broadly that growing food should be the most natural of endeavors, fostering diversity and balance, which are the signature of healthy communities.

From northern Alberta to California's Central Valley and beyond to wherever insect-pollinated crops grow, the global pollinator crisis can inspire us to rethink the most fundamental challenge of modern agriculture: How can we grow sufficient food to feed a massive human population while at the same time protecting the environment?

Pollinators teach us that the answer may be to manage less. Instead, we might conserve and restore the habitats that provide economic services and the balance that is so critical to human contentment, the sense that we are living comfortably and in harmony with the nature around us.

At one point in my California trip Claire Kremen and I were strolling along a tiny hedgerow at the edge of a vast sea of almond trees. I asked her what she would consider success in a revised version of agriculture.

She looked out on the fields for a few minutes, lost in an imagined vista of a Central Valley revolving around ecosystem services, before replying: "I think the landscape would look really different."

6

Bees in the City

It was tobacco that brought Alice and Désirée to my bee research laboratory in 1999 for advice on their eleventh-grade science project, "*Nicotiana tabacum*: Not Only Smoke!" for the Vancouver Regional Science Fair.

Alice and Désirée were from Milan, Italy, and their families were tight enough that when Alice's family immigrated to Canada, Désirée's soon followed. Although the girls had quite different personalities, they seemed closer than sisters and were most definitely a team. Alice was high energy, talkative, and bursting with plans and enthusiasm. Désirée was quiet, verging on shy, and contributed a steadiness and an organizational capacity that balanced Alice's wackiness.

Their science fair project had nothing to do with bees but rather with developing nonsmoking uses for tobacco to aid farmers as antismoking campaigns reduced the market for their crop. They had met one of my graduate students, who offered to provide some advice on their submission, and happened to come into the lab on a day when a visiting scientist from Utah was giving a talk they attended on wild bees in urban habitats.

They burst into my office immediately after I returned from the lecture, Alice talking nonstop and both of them brimming with inspiration. "We want to do that," she said, "study wild bees in Vancouver." I read their enthusiasm as momentary and certainly not realistic as we had no funds for wild bee research. To get them to retreat back out the office door, I gruffly pointed out that they would need a fairly large grant, and if they were interested, they needed to write one first.

I assumed that would be the end of it, but, oh, was I wrong. They returned two days later with a highly imaginative proposal, the "Once upon a Bee" project. It began with a fairy tale, illustrated with their own cartoonlike drawings: "Once upon a bee, when the city of Vancouver didn't exist, small simple animals like bees, hummingbirds, and flies played a crucial role in preserving the greatest treasure of planet Earth, biodiversity." It then jumped ahead in time to the founding and expansion of Vancouver: "Trees were cut, meadows and forests covered by concrete, and bogs filled. The rare leftover patches of wildlife became smaller and smaller until they were so small that they couldn't provide enough food for native creatures to survive. The bees started to disappear . . ."

They didn't stop with a fairy tale, however, but went on to propose a substantial and well-conceived project that included studying the diversity and abundance of wild bees in the city of Vancouver, as well as developing a bee-conservation plan for the city and a broad spectrum of educational programs for schoolchildren, garden clubs, and our local science museum.

I'd never seen a grant proposal quite like it. For one thing, no one begins grant proposals for scientific research with stories, let alone cartoons and a fairy tale. Yet the quality of their proposal, in terms of their writing and their research plan, wasn't much different from those I'd seen from graduate students. Their hypothesis that bee populations would be diminished in cities was highly testable through the data they would gather, and their vision of a comprehensive research, conservation, and education project was stunningly appealing.

Unable to resist, I promised them to work a bit on the grant to smooth out the rough edges and submit it to some private foundations that might be intrigued by the project. To my great surprise, we raised $70,000 from two foundations, and the Once upon a Bee project was a go.

And the science fair project on tobacco? They won the gold medal that year.

o o o

Nature can be found in unexpected places, even cities. Because they are small and obscure residents, bees escape our notice. Yet a closer look for bees in the city enhances our appreciation for the beauty, diversity, and significant role of urban nature and reveals their critical function in maintaining green city environments through pollination.

Cities are inhabited by many subcultures of creatures that escape our attention. Our human city moves at a bustling pace inconsistent with noticing other organisms. Our urban search images are for shops and restaurants, entertainment, and housing but not so much for nature.

Cities are highly disturbed, complex habitats that attract an eclectic mishmash of species present for often-quirky reasons. They contain much nature that's subtly below our radar, with entire ecosystems thriving in and around roads, malls, offices, and homes.

That's where Alice and Désirée went in search of the feral bees. They collected them from a great range of habitats, including community and botanical gardens, urban wild areas, backyards, railroad rights-of-way, and road edges. In each broad habitat they mapped out a number of small plots of twenty-five square meters each to collect from so that they could compare numbers of bees in the various habitats.

It's quite an art, collecting bees. First you need to establish a search image for bees as they dart between flowers and their nests. Otherwise, you'll collect an inordinate number of flies and other nonbee creatures. Bees have a distinctive way of flying, and experienced and discerning entomologists develop a sixth sense that triggers them to pounce when a beelike blur whips by the corner of their eye.

Nets are the bug collectors' instrument of choice, and Alice and Désirée became acrobats with their nets, whipping them around with precision and verve to trap the quickly flying bees. They spent most summer days seeking the wild bees, and by September they had accumulated a thorough sampling of Vancouver's bee world. They then spent well over a year in the laboratory laboriously keying out each bee to species, often differentiating between species by the tiniest structural parts, and then sending some off to specialists to confirm their identifications.

To our surprise, they collected 2,600 wild bees and 911 honeybees from just the small plots they used for sampling, anywhere from 2 to 32 bees per hour. Their collections included fifty-six species, with bumblebees being the most common, even more abundant than managed honeybees. The girls also determined the bees' nesting sites, mostly in cavities, pith, soft wood, firm wood, or soil, and noted the flowers they visited, most prominently dandelion, cotoneaster, and blackberry.

This high diversity and abundance of wild urban bees compared to honeybees made it clear that wild bees are important pollinators for backyard fruit trees and gardens, natural vegetation, urban parks, and the in-between zones where nature intercalates into the city. But were our Vancouver numbers high or low in comparison to wild bees in other cities?

Turns out we're on the low end, possibly because of Vancouver's notoriously rainy climate, which is too wet for many bee species. The world record for wild urban bees is held by Berlin, where 262 species are found, mostly from disturbed areas, rural patches, road edges, and backyard gardens. Next in line are the five boroughs of New York City, with slightly more than 200 species. Sao Paolo, Brazil, is midrange, with 133 species, while Boulder, Colorado, is home to 88, and Albany and Berkeley, California, have 74.

It's comforting to know the wild bee subculture does well in urban habitats, but it also says something about nature in cities generally. Urban environments are humanized by well-designed parks, golf courses, homeowner-planted flowerbeds, backyard gardens, and botanical reserves interspersed between pavement, buildings, and weed-infested empty lots and rights-of-way.

Most of the animals that populate our cities are generalists that do well in this type of disturbed habitat, including rats, deer, pigeons, coyotes, beavers, and geese. These nonspecialists thrive in a wide spectrum of situations and have become so abundant in cities that they are considered pests.

Urban bees are similar in that it's the generalist bees like bumblebees that thrive, but they differ in not being pests. Rather, bees provide a highly beneficial ecological service: pollination.

Moreover, Alice and Désirée's research revealed that bees within Vancouver were about as diverse and abundant as bees in nearby agricultural natural habitats, but the species composition was different. Our "Once upon a Bee" hypothesis, that bees were disappearing in the city, was wrong. It's not that bee numbers were low in Vancouver, just that they differed from the bees outside urban areas.

Recently, planners have recognized that cities don't need to be barren landscapes and are encouraging more coherent and locally appropriate habitats. The trend has been toward wildlife corridors, unmanaged green spaces, and backyard plantings with native rather than introduced species. And there's another trend in urban planning: growing food in cities. Bees, both wild bees and managed honeybees, are turning out to be a core urban resource for this expanding movement to bring agriculture back into cities. Consideration of bees in cities reveals how habitats can be designed to meet both natural and human needs.

o o o

City farming is exploding. Peter Ladner, former city councilor and publisher of the magazine *Business in Vancouver*, described the growing movement in his book, *The Urban Food Revolution:* "Community gardens are popping up everywhere, and newbie gardeners are lining up for plots. On rooftops in cities all over North America, people are tending beehives, growing herbs, and supplying leafy greens to restaurants. Planners, food policy councils, legislators, and urban conferences are diving into the local food pool."

This urban food revolution is actually more of a reinvention than a revolution, as urban agriculture is already a significant element of global food production outside of North America. For example, 45 percent of all vegetables in Hong Kong and 80 percent in Hanoi are grown on urban farms and gardens. Remarkably, 80 percent of Singapore's poultry and 90 percent of the eggs and all of the milk in Shanghai come from within those cities. In contrast, cities in the United States grow less than 10 percent of their own food, although during World War II 40 percent of all American produce came from home gardens.

Why urban food, and why now? Ladner points to a number of factors that emerge from many global surveys, including the stimulation of local economies and an urban interest in supporting family farms. Urban food is also perceived as healthier as it's not genetically modified and is usually grown with fewer pesticides. Finally, urban agriculture is considered to be environmentally friendly as it preserves green spaces and cuts down on energy use by reducing shipping of food to global markets.

I happen to live in the heart of one of the world's most progressive and environmentally sensitive cities, with municipal policies encouraging urban agriculture. Vancouver is full of community gardens, more than seventy-five of them, because our city government encourages residents to grow food on vacant park, school, and other land, including the front lawn of city hall.

The city also allows developers to reclassify a potential building site as a public park or garden during the many years it takes to begin a major construction project. Because city taxes are 80 percent lower with that classification, developers often save hundreds of thousands of dollars. Then, when the developments are built, they are required to include edible landscaping like fruit trees and berry bushes, as well as food-producing garden plots on rooftops and in courtyards.

Residents also have been proactive about growing food. Homeowners in one Vancouver neighborhood banded together to form a "two-block diet garden." They share tools, a beehive, chicken coops, a greenhouse, compost, seed purchases, and information, jointly harvesting and sharing the yield from each of their private yards.

Other residents provide land in their yards for professional urban farmers who will do all the work, pooling the produce from a collection of front and backyard minifarms and redistributing to the landowners for a fee. One former lawn-care franchise owner in Vancouver switched his business entirely to backyard food harvesting, returning food to homeowners for the same payment he used to charge for lawn care.

Bees are a necessary element if this burgeoning urban agriculture is to be successful since the large majority of fruits and vegetables grown in cities require pollination by either wild bees or honeybees. It's a pretty basic relationship: no bees, no food.

With this in mind, Alice and Désirée recognized the value of educating Vancouverites about wild bees and were dynamos around town for years in promoting bee-friendly practices. They put on programs at our local science center, gave innumerable talks to school groups, and were popular guests at garden clubs and ecologically focused public groups.

The Once upon a Bee project was probably the first extensive urban education and conservation project that focused on wild bees, but similar and considerably larger projects have sprung up all over the world. The Great Sunflower Project is the queen of bee surveys and conservation programs at the moment and may be the largest citizen-based science project ever conducted. Founded and run by San Francisco State University biologist Gretchen LeBuhn, it has organized many thousands of residents all across the United States to observe bees and feed data into her national database.

"Sunflower" refers to the flower of choice for participants to observe bees, although there's a short list of other flowers. Observers are asked to count the number of bees on target flowers in fifteen minutes, and once a year, on National Bee Day, they watch for an hour on the designated Saturday, from 9:00 to 10:00 a.m.

Massive amounts of data have resulted, providing an ongoing snapshot of bee abundance locally and nationally. Currently the average number of bees observed in US cities is twenty-three per hour, a benchmark against which to measure the success of conservation campaigns or the extent to which bees are threatened in the future. These data can also be helpful in determining where pollinator service is strong or weak compared to averages.

A similar partner project emerged in New York City in 2007, the Great Pollinator Project, launched by the American Museum of Natural History and the Greenbelt Native Plant Center. Volunteer bee watchers observed the bees visiting selected flowering plant species such as coneflower, mountain mint, and rough-leaved goldenrod and assigned bees to one of five categories: honeybee, bumblebee, large carpenter bee, shiny green bee, and other type of bees. More than fifteen hundred observations have provided an extensive profile of wild bee populations in the city's five boroughs.

In the United Kingdom, the Urban Pollinators Project planted flower meadows in four cities, Bristol, Reading, Leeds, and Edinburgh, to determine whether such plantings increase the diversity and abundance of wild bees. The German government recently released a bee app with information on more than one hundred bee-friendly plants and is encouraging urban architects and planners to consider the needs of wild bees in their development plans.

A number of consistent messages emerged from Alice and Désirée's project and the many others run by bee-loving citizens and scientists from cities around the globe. Wild bee conservation basically requires three elements, beginning with nesting sites.

Wild bees are diverse, and each species has slightly different nesting requirements. To cover the range of possibilities, Alice and Désirée suggested maintaining areas of loosely packed, well-drained, and mostly bare soil, providing logs in various stages of decay for the wood-nesting bees, and leaving bundles of brush for stem and twig nesters. Artificial nest boxes drilled with hollow holes of varying sizes can supplement more natural sites.

The second requirement is for nectar- and pollen-producing flowers. Plantings of fruit, vegetable, and berry crops as well as ornamentals are important, particularly native plants as they are most likely to attract native bees. It's also useful to leave waste areas like roadsides, railroad rights-of-way, and empty lots unmanaged, as weeds can be excellent food sources for bees.

Finally, bees are sensitive to many pesticides, and it's critical to reduce chemical use in the city. Insecticides are problematic as they often kill nontarget insects like bees, but weedkillers are also an issue. It's not that herbicides kill bees, but they kill the weeds that bees forage on. A lawn full of dandelions is bee heaven, and bee conservation benefits from homeowners who favor bees over immaculate lawns.

It's not only wild bees that have been part of this back-to-the-land urban crusade. Honeybee colonies are expanding in cities all over the globe—partly for the honey they produce, partly for their benefits as pollinators, but also because urbanites crave community and connection. Beekeepers are as social as their bees, and beekeeping provides an excellent opportunity to socialize with other keepers and to meet your neighbors over a friendly jar of honey.

Honeybees have become an urban trend, from the rooftops of five-star hotels to backyard and community gardens to the grittiest, most poverty-stricken city corners. In the process, beekeeping serves as a bridge between the most disadvantaged and the prosperous, spanning income gaps with a common passion.

o o o

It's a small zone of urban tranquility, a rooftop garden twenty-one hundred square feet in size and located on a third-floor hotel terrace. It was one of Vancouver's first green roofs, now growing sixty varieties of herbs, vegetables, fruits, and edible blossoms, pollinated by a few beehives tucked away in a garden corner.

Not just a random isolated roof, this represents a significant trend in the hospitality industry. Gardens and the bees that pollinate them are found on many of the most exclusive hotel roofs in the world.

Vancouver's first hotel eco-rooftop is about twenty years old, although the bees are more recent. It's in the absolute heart of urban downtown, at the Fairmont Waterfront, overlooking the harbor on one side and tall skyscrapers in other directions. Similar rooftop beekeeping can be found at Fairmont hotels across Canada as well as in San Francisco; Washington, DC; Newport Beach; Dallas; Seattle; and Boston. Many other four- and five-star properties are doing the same, including New York's historic Waldorf-Astoria hotel, Paris's Eiffel Tower Hotel, and London's Royal Lancaster.

The Fairmont widely promotes bees as part of its eco-hotel brand. The bees on their roof produce more than six hundred pounds of honey a year, and the hotel maintains an additional nineteen colonies near the Vancouver airport that bring in another forty-two hundred pounds. Each of their rooftop apple trees went from twenty to two hundred apples when honeybees were added to the garden, providing enough yield to keep the restaurant supplied in ingredients that are as local as it gets.

The Fairmont's executive chef, Dana Hauser, views honey as fitting into both the hotel's signature cuisine and its commitment to local ingredients and ecology. She's originally from Upper Island Cove, Newfoundland, and still talks with a lilting accent from back home. She said, "I like to really support our local community; we're focused on the one-hundred-mile lifestyle, using food from nearby as much as possible. The Fairmont Waterfront is committed to being a green hotel . . .

[Bees are] our way of encouraging the pollination of flowers in our area."

Her signature honey dishes include organic Fraser Valley mesclun greens with rooftop honey lavender vinaigrette, rooftop honey-roasted breast of duck, and warm apple pie with honey goat cheese ice cream.

The hotel offers complimentary tours of the garden, including the unique experience of tasting honey right from the hive. As Hauser described it: "You've never tasted honey until you've tasted honey that comes directly from a frame. You pull the frames out, they're dripping everywhere. It's a sensory experience. They spoon the honey up from the frames and can taste the just-made fresh honey, still warm, and chew the wax—they love it. And I love that last lick of the spoon. That's the kind of stuff you come to work for every day."

The Fairmont was Vancouver's first hotel rooftop location for beehives, but there were earlier sites motivated less by the green movement and more by using rooftops to hide colonies from city authorities enforcing bylaws prohibiting beekeeping. There's a famous story in Vancouver of a crusty old-time beekeeper who, decades ago, kept twenty colonies on the roof of a building in Chinatown, painting them black for camouflage. He bragged about his incredible yields of four hundred pounds per hive until someone pointed out that his bees were only half a mile from a large sugar-processing plant. The "flowers" they were foraging on were piles of white sugar waiting to be loaded onto ships in the harbor.

The surreptitious nature of Vancouver beekeeping shifted in 2005, when the city changed its bylaws to allow beekeeping. Colonies and their keepers came out of hiding. One in particular, Allen Garr, emerged unofficially as the senior figure passing on lore, training new beekeepers, and managing the more high-profile colonies in public gardens, at Science World, and on the Convention Centre's green roof.

Garr is in his early seventies but looks at least ten years younger. He's a journalist by trade, working at times for the Canadian Broadcasting Corporation's national television news and at others as a writer for local Vancouver newspapers. He's thin, fit, wears a simple gold earring in his left ear, and has the sweetest of dispositions. And he's passionate about raising bees in the city.

I talked with him at length in his 1920s-built Vancouver home. It sits in an older neighborhood lined with leafy trees and filled with character homes from that era. Garr's house has a large, welcoming porch, the original fir floors, stained-glass windows, sliding hardwood pocket doors, and that quirky, patched-together architecture that results from numerous additions and renovations.

We met in his basement honey house, where chats with beekeepers anywhere in the world usually start. Urban honey houses tend to be small and efficient; Garr's is in a corner converted for extracting honey and building and storing equipment. It has that signature beekeeping smell of wax and honey and is decorated in beekeeper funk with used overalls, boxes, lids, and veils stacked all around the room. When I arrived, he was washing out his honey-extracting equipment in the basement sink, marking the end of this year's fall extracting season.

I asked him how he got his start in bees, and it turns out he was reluctant: "A friend of mine was a beekeeper. He asked me if I was interested in keeping bees, I said I wasn't, then he asked me again and I said I wasn't, and then he picked up a swarm on the west side of town, dropped it off on my back deck, and said, 'You're a beekeeper.'"

Beekeeping was still illegal when he started, around 1974, and it appealed to his rebellious side: "I'm an old hippie. Keeping bees in the city illegally was kind of cool, like having a grow-op."

There's an urban legend about Garr that I had heard many times but doubted: that he was allergic to bee stings. It turns out to be true. He spent seven years getting desensitizing shots, which reduced the danger to some extent, although he still winds up in the emergency room on occasion after a particularly effective sting.

I asked him why he would maintain a potentially life-threatening hobby, and like most beekeepers he has multiple reasons behind his passion: "I like animals. When I was a kid, I had tropical fish and pigeons I'd breed and show at the Canadian National Exhibition. Bees are just really fascinating, the way they work as a collective. I also like the impact they have on the environment. It's plugged me into the world, a prism through which I learned to be an environmentalist."

Urban beekeeping has increased exponentially in recent years in spite of colony collapse disorder and the myriad pests and diseases facing contemporary beekeepers. Where ten or so older men used to show up at a typical Vancouver bee club meeting, average attendance is now more than 110 and includes many women and younger members.

And it's not only Vancouver. In Toronto, sales of beekeeping equipment hit a record high this year, and registrations for a community-run introductory beekeeping course have jumped from a handful of participants each year to more than one hundred. In London, England, hive numbers jumped from sixteen hundred to thirty-three hundred in the last four years.

France has become another center of urban beekeeping, with the French government encouraging honeybees in cities throughout the country. Paris is particularly honeybee friendly, with rooftop hives popping up on luxury hotels and skyscrapers and in the renowned Luxembourg Gardens. And honey from city bees has become fashionable, with the finest restaurants rolling out signature honey-based cuisine.

Perhaps the most famous urban beehives are atop the Paris Opera house, each producing between 110 and 180 pounds of honey annually. Beekeeper Jean Paucton points to the ban of pesticides in Paris and a wide variety of flowers as factors contributing to consistently higher honey yields in the city than from hives kept in the surrounding countryside. An analysis of pollen in Parisian honey showed more than 250 different floral sources compared to around fifteen to twenty in country-produced honey, supporting Paucton's belief in the richness of Parisian habitats.

Garr attributes some of the growth in Vancouver to beekeeping going legal: "The bylaw made a big difference—not everyone wants to do things that are illegal." Beekeeping as an urban hobby also has the advantage of requiring only a small space, even just a balcony or tiny backyard.

He also noted some deeper reasons behind beekeeping's current urban popularity that place bees as an essential element of an increasing sense among urbanites of connecting to the environment: "People are freaked out about the environment and global warming and want to do something. They're worried, trying to have a healthier world. It's the same people who used to go back to the land in my generation. They're staying in the city and buying a place where they can tear up a lawn and put in a garden and beehives. The same people who are involved with urban agriculture are into beekeeping."

Besides their environmental and agricultural mindsets, urban beekeepers possess a streak of oddball; it's an edgier hobby than, say, knitting scarves or growing zucchini. It also attracts the exceptionally curious, those eager to learn and share information. New beekeepers quickly become rabid attendees of bee club events, talks, workshops, and field days.

What's more, urban beekeeping is not only for prosperous middle-class residents. Bees are also being used in tougher settings to provide livelihood and deliver social services to city residents who have addictions, mental illness, or disabilities or are homeless or poverty stricken.

o   o   o

Hives for Humanity began with one hive in Vancouver's downtown eastside, the poorest postal code in Canada. It was founded by mother-daughter duo Julia and Sarah Common. Julia's the mom and the beekeeper, while Sarah is a social worker. Together they provide skills training through Vancouver's Portland Hotel Society, a service agency that delivers housing and other support to individuals who are hard to house and at risk of homelessness due to substance dependencies and challenges in their physical and mental health.

Julia and Sarah brought honeybees into this difficult neighborhood. Their hive was kept last year in a small garden a few blocks away, next to Insite, the safe-injection site that has garnered considerable international attention by providing a supervised and legal facility for intravenous street drug users to shoot up.

Walking to Sarah's office in the society's Drug Users Resource Centre is quite a different experience from strolling through Garr's neighborhood or visiting the Fairmont's terrace beekeeping. I passed through block after block of soup kitchens, social service agencies, addiction-treatment centers, and single-room-occupancy hotels. Yet, the streets are lively with community members socializing, and there is a sense of possibility amid the poverty and addiction.

I talked with Sarah and one of her assistants from the community, Yvonne Yanciw, who lives next door to the Resource Centre. Yvonne has had some difficult times in her life, and working with the bees has been an important part of her recovery. When I arrived, three or four beginning beekeepers were hammering together hives under a sign that said "Believe Life Hope," so we moved to a quieter office to chat.

The introduction of beekeeping to this impoverished area is part of a larger community movement to grow healthy local food as an essential service. The bees have been a particular source of pride, providing training for residents and employment.

The pilot hive produced 120 pounds of honey that sold out quickly and generated additional income through the crafting and sale of beeswax candles. Hives for Humanity is ambitious to improve poor communities across Canada hive by hive. The program expanded to seventy colonies in 2013 and is seeking significant funding to realize its astute business model as a social enterprise venture.

Economic development is an important objective, but the therapeutic aspects of beekeeping may be Hives for Humanity's most significant contribution. I asked Sarah and Yvonne about the healing aspects of the program, and Sarah noted that "It's really therapeutic to be in a community [of] people all caring about something, having ownership, building self-worth and pride. To look ahead—that's something that's rare in a community often absorbed in a lot of chaos and trauma and needs of the moment."

Being one of the beekeepers has had a great impact on Yvonne, calming her personally and providing a source of self-confidence: "I'm a real high-energy person, but I have no problem being slow and steady and calm in there. You can concentrate." An astute observer, she has noticed how being around the bees has affected other community residents: "Life down here sometimes has snags in it; the bees give them a chance to concentrate on something busier than they are, gives them a chance to slow down."

This year Yvonne is getting her own hive, which she is going to manage on the rooftop of her housing complex: "It's something I can do. Most people here just live for the day. The bees provide a future, and not many people look to a future down here."

Another massive project that connects bees and poverty is Growing Power, an urban farming collective founded by former professional basketball player Will Allen. This community-driven enterprise began in Milwaukee about twenty years ago and then expanded to Chicago, its mission to provide healthy and affordable food for inner-city residents.

Its overall objective is to create holistic and organic community food systems through training and hands-on experiences. It's been highly successful, growing enough food to feed ten thousand city dwellers each year. Many hundreds of volunteers contribute labor at the numerous urban sites while learning about farming and nutrition.

Beekeeping has been a crucial element of Growing Power since its inception. The project currently maintains fourteen active hives in Milwaukee and six in Chicago, producing about 150 pounds of honey per hive each year and pollinating its extensive vegetable gardens, berry patches, and fruit orchards. In Chicago, the project's Youth Corps harvests beeswax to process into lip balm, soap, body scrubs, and candles.

In Saint Louis, Missouri, honeybees are being used to expose disadvantaged youth to entrepreneurship and sustainable living through the Sweet Sensations project. The young people learn work and life skills by planning and taking care of an apiary, marketing honey and beeswax, developing business plans, keeping financial books, and planning strategically to grow their business.

From these and many other projects it's apparent that beekeeping with honeybees, combined with efforts to value and enhance wild bees, have become a significant aspect of urban culture. Whether for environmental, agricultural, or antipoverty reasons, bees are now recognized as integral components of city habitats.

o o o

Underlying the stories of bees in the city are the profound benefits derived when we successfully weave the natural and the managed worlds together. Bees in the city provide a template upon which to reflect on how to reconcile our human needs with those of the other species with which we share our planet and on which we are dependent in so many ways.

Natural and unmanaged versus artificial and managed is not a stark divide, just as "city to wilderness" is a continuum rather than a border. Some say we should leave a minimal footprint on the earth, but it's more realistic to hope for a footprint light enough for other creatures to thrive while heavy enough to meet human needs for food, energy, and shelter. Balance is the desired end point, and that's not always easy to achieve.

Wild bees and honeybees in the city provide an interesting example of how the natural and the artificial worlds can be compatible and thrive together. Bees fit well in the interstitial space between disturbed and natural habitats, thriving so long as forage, nesting sites, and a toxic-chemical-free habitat are available.

These elements are all possible in urban settings. Cities that have proactively introduced policies and programs to enhance bee-friendly elements have found that both environmental and agricultural interests are well served.

Bees in cities also express a subtle tension between wild bee proponents and beekeepers since excessive numbers of honeybee colonies could competitively exclude wild bees. In many ways wild bees and honeybees have the same requirements and serve as an appropriate model to examine how best to navigate conflicts between nature and humans.

Creating pollinator-friendly habitats is good for all pollinators, and most experts don't think we have yet saturated cities with honeybees to the point that wild bees are being competitively excluded, due in part to progressive municipal programs. Improving habitats to provide room for both the wild and the managed is an elegant approach to achieving balance.

Our challenge is to enhance habitats sufficiently to sustain both human endeavors and native species. Bees demonstrate that ecologically coherent cities and urban agriculture can be compatible and suggest that a similarly balanced approach might also work well in rural ecosystems and farmland.

Urban habitat enhancements such as those suggested by Alice and Désirée expand opportunities for wild bees, while honeybees in the city encourage plantings attractive to both managed and unmanaged bees. Shifting urban nature toward more indigenous local species creates more ecologically coherent urban flora and fauna. Because honeybees were imported, the enhancement of native wild bees is particularly important in restoring some balance between invasive and indigenous species.

I'm currently living in an apartment in the heart of downtown Vancouver but previously lived for twenty-three years in nearby New Westminster in a suburban, single-family neighborhood. When I moved in, there was an elderly gentleman across the street with two hives on his flat garage roof, which he accessed by climbing out of an adjacent bedroom window. Next door to him was a schoolteacher who kept ten colonies in his backyard.

I added my two hives, and like all beekeepers we traded honey and stories, helped each other catch swarms, and complained about the weather. I'm sure we had well-pollinated backyards for a mile or two in every direction. My two hives produced the best honey crops of all my two hundred or so colonies, most of which were outside the city in rural farmland.

I was equally attentive to Alice and Désirée's advice about wild bees. I kept logs, soil, and brush around for the wild bees to nest in, planted bee-friendly plants, and didn't mow when dandelions and clover were blooming. In that way my small backyard was a minimodel for urban bees, enhancing both the wild and the managed side by side.

It was a small space but expressed a much broader philosophy about our human place in nature. The wild and the managed were equally promoted, and the presence of both increased my appreciation for that balance point, the sweet spot where nature on the one hand and human ingenuity and commerce on the other are in equilibrium.

Bees can help us not only bridge the natural-to-managed divide but also straddle another deep division in North American society, rich versus poor. I was struck with how easily bees took me through diverse social and economic strata, from exclusive high-end hotels to comfortable middle-class neighborhoods to the most extreme poverty and social chaos.

By cutting across class and income, interest in bees is a great equalizer. Bee enthusiasts are united by a passion for maintaining ecologically healthy habitats, an interest in protecting the agricultural benefits bees provide through pollination, a fascination for the behavior of all bees, whether wild or managed, and respect for the hard work bees put into producing the honey and wax that beekeepers harvest.

Bees also help resolve another conundrum for urban dwellers, the sense of isolation and lack of community that often afflicts cities. Bees connect residents to their neighbors, to other bee enthusiasts, and to the growing movement of urban gardeners and environmentalists. Common interests like bees build microcommunities that transform urban loneliness into that sense of belonging so fundamental to human happiness.

That's the great lesson we can learn from urban bees. By pollinating flowers, bees literally connect one generation of plants to the next and guarantee the continuity of a healthy urban ecology. We, too, need to consider ourselves as connectors, part of a larger chain that passes on values and a sense of belonging to each other and to the generations to come.

Bees have a way of doing that, reminding us that it's only together with nature that we can guarantee our own prosperity and survival even in humanity's largest hive, the city.

7

There's Something Bigger than Phil

There's a comedy routine from 1961, "The 2000 Year Old Man," in which Carl Reiner interviews a two-thousand-year-old Mel Brooks. At one point Reiner asks Brooks how humans discovered God. Brooks replies in his best old-man accent: "Well, even before the All Mighty, we believed in a superior being. His name was Phil. Out of respect we called him Philip. Philip was big and strong; nobody was as powerful as Phil. If he wanted, he could kill you. As a result, we revered him and prayed to him: 'Oooh, Philip, please don't hurt us! Philip, please don't pinch us!' But one day Phil was hit by a bolt of lightning. All of a sudden we looked up in the sky and said, 'There's something bigger than Phil!'"

Hilarious when performed by two exceptionally funny guys, but it's also the most serious of human queries: Is there something bigger than us? We may probe this deepest of issues through religion, spirituality, contemplative practice, science, art, or philosophy, but underlying every human reflection is our compelling need to understand who we are and why we are here, to consider whether the universe around us has meaning.

Given the power of these universal questions it's not surprising that our thoughts about bees have expanded beyond the practical aspects of beekeeping into wonder and fascination and even further into imbuing bees with powerful qualities and mystical powers. Bees have been connected to human spirituality since the earliest recorded history and are deeply embedded as symbols and stories in mythology, religion, morality, ethics, and philosophy.

Honeybees are also associated with health, the products of the hive renowned in folk wisdom for their healing properties. Pollen, honey, royal jelly, and even stings are applied for myriad ailments, although there has been little rigorous testing to confirm their efficacy.

Practitioners who use bee products and those who incorporate bees into their worship believe that there is more to bees than scientific knowledge about their biology or management would suggest. It is faith rather than data that rules in both spheres, with apitherapy and spirituality found in the realm of true belief, challenging us to reflect on how we "know" things.

Why are we so mesmerized by bees? We turn to them as guides to the eternal mysteries, use them as examples of desirable human virtues, and apply their products for our physical and spiritual healing. Do bees just provide useful metaphors, or do they actually possess qualities that hold the key to profound understandings and health?

o   o   o

There are many approaches to comprehending the unknowable, ranging from mainstream religion to the most marginal cult, philosophy to science, art to music. They each have their leaders, but shamans are particularly intriguing as intermediaries who connect us to the mystical. They work their magic through reaching altered states of consciousness and interacting with the spirit world, a dimension nebulous to most of us but physical and real to shamans.

Their experiences, often induced by drugs, seem supernatural. Shamans frequently describe travel to other worlds, dimensions, and times, and it's tempting to dismiss their experiences as outside reality. But the popularity of someone like Carlos Castañeda exposes a deep urge we have to cross the border between the truth we see and that which we might imagine.

Simon Buxton is the Carlos Castañeda of bee shamanism, an elder in the hierarchy of practitioners referred to as the Path of Pollen. He is founder and director of the Sacred Trust, a UK-based organization dedicated to teaching practical shamanism for the modern world.

The Way of the Bee includes every new-age realm of the spiritual: healing, past lives, sacred sexuality, occult rituals, and ethereal beings that take corporeal form. It's difficult to discern whether its followers believe that their experiences while in sting-induced trances are real or should be considered symbolic.

Buxton's entry point to bee shamanism was at the age of nine, when a neighbor, an Austrian bee shaman, cured him of encephalitis. In his book, *The Shamanic Way of the Bee*, he recalls looking into the eyes of his friend, whom he calls Herr Professor, and seeing "countless magnificent hexagonal lenses, each one of them able to see deep into my soul." A few days later he's cured.

His next teacher on the Path of Pollen was a beekeeper nicknamed Bridge, whose full name was Bid Ben Bid Bont, a name given to him by the bees that means "who would be a leader, must be a bridge." Bridge "lived simultaneously in the past, the present, and the future, a bridge across, through, and outside the circles of time." Those who complete the path develop the ability to transmute matter, heal disease, and prolong the human life span.

Buxton's training began with appreciation for the intricate biology and relationships within the hive and between bees and flowers but soon progressed into the occult. His first serious ritual involved being stung multiple times by his spirit guide, Bridge, on particularly significant acupuncture points, which when "stimulated by the Sacramental Venom, allows the initiate to enter in the words that exist outside of time and space."

In a trance Buxton enters the hive, merges with the colony mind, transforms into a drone, and mates in passionate embrace with the queen bee. In a later trance he meets up with the Queen of Synchronicity and her six attendants in human form and from them learns about the mystical powers of the six-sided hexagon *(hexagramma mysticum),* the perfect proportions that allow for harmonious growth and development.

The queen also unveils the mysteries of the figure eight *(lemniscus infinitorum),* which is the dance pattern bees use to communicate the distance and direction to flowers. This pattern is a key to sexual union, and Buxton ends up copulating with one of the queen's six assistants once he is appropriately indoctrinated into the *lemniscus.* It also is a vehicle through which elders in the Way of the Bee travel to other times and worlds.

I found the Path of Pollen to be pretty far from my conventional entomological training and decided I needed my own spirit guide to provide more explanation. Coincidentally, or perhaps cosmically, a Facebook friend of mine shared a message about sacred beekeeping posted by a local Vancouver woman, Nikiah Seeds, the same day I discovered the Way of the Bee.

Seeds, who describes herself as a sacred beekeeper, teaches at the Ashland, Oregon, College of the Melissae/Center for Sacred Beekeeping. I arranged to meet at her Vancouver home, which appropriately has nesting boxes for orchard mason bees at the front gate and a hive of honeybees in the backyard, next to her chicken coop. The honeybees were at the center of her garden, which is laid out like a medicine wheel, symbolic of indigenous North American culture.

We met in her light and airy kitchen, with the vague smell of incense permeating the air. Seeds is lively and enthusiastic, trained as a priestess in goddess- and earth-based traditions. She's worked as a doula, coauthored a book, *Reclaiming Women's Menstrual Wisdom,* and marketed a line of herbs and spirit medicine teas for pregnant women. She prefers the term "pathfinder" to shaman and struck me as unexpectedly well grounded for someone so immersed in the occult.

Similar to Buxton, she has "been taken into the hive by a bee" and visited with ancestral spirits. She believes her experiences aren't real but imaginative, which I found reassuring. She uses a drum, not drugs or stings, to induce a trance state and is then able to pull up mental images of places and spirit guides that range from her own Ukrainian beekeeping ancestors to bees themselves.

I asked her about what happens on these spiritual journeys: "I have been learning from the hive consciousness, the bees have been asking me over and over to listen to their teachings and in turn asking me to share their wisdom, their honey, and healings."

She's heard three clear messages from the bees, the first being that they are suffering from colony collapse disorder and that we humans need to pay more attention to the environment they—and we—live in: "The bees are desperately trying to get our attention to support the environment. There is so much we lose without them."

The second message has been more personal: to "Calm myself, still myself. You have to be grounded to work in a bee yard. You can't go in there harried and rushed; you'll get stung. Take a deep breath and calm down." Her third lesson has been about the importance of community: "A hive isn't a set of individual bees or the queen. The entire hive is a consciousness of one collective group. You would never go and work with one individual bee. Community is a big part of the bees' lessons."

Bee shamanism can easily be dismissed as eccentric, but is the Path of Pollen so different from any of the tools we use to probe the unknowable? We humans have always used the world around us to cast our questions about meaning and to divine the future.

We have imagined meaning in the patterns of stars and created gods associated with the heavens, the waters, the land, and the world below ground. We've imagined relationships with plant and animal spirits to hear what they could tell us about hunting, gathering, and the origins of Earth and its living species. We pay attention to the interpreters of spirits and gods, including shamans and witches, holy men and women, and conventional priests, rabbis, and ministers.

Seeds uses an approach that is rooted in traditions most of us don't follow or fully comprehend, but the more we talked, the more I could see that we aren't that different in how we think of bees as an inspiration, a guide in deepening our commitment to a healthy environment. Beekeepers will tell you they grew as environmentalists by noticing how their bees depend on flowers and the flowers on bees, although the language they use to describe that experience has more in common with farmers than shamans.

Further, beekeeping requires an awareness and intuition of what is going on in a hive. Beekeepers tap into that hive consciousness as Seeds does, using all of their senses to discern how the colony is functioning and to determine whether management action is needed as a way of collaborating with the bees to improve the colony's well being. "Awareness" is a more palatable term than "hive consciousness" for most beekeepers, but essentially they mean the same thing. Bees can be spirit guides whether we see ourselves mystically or practically. Seeds put it well: "You don't have to be a priestess or a shaman to treat the bees respectfully. The whole aspect of beekeeping is a spiritual practice. I find it so deeply beautiful; it's a special calling. I think it's pretty sacred, I think it's pretty special. How could I not tap into what the bees have to offer us?"

o   o   o

References to bees and honey are not only for those on the edges of sacred practice but are also found as commentary in mainstream religions and abundantly referenced in secular life. Bees are depicted in religion and culture more as metaphor than through ecstatic experience, with observations about bees providing lessons for how to live our human lives.

Honeybees are frequently held up as examples of hard work and diligence. Mormons have been particularly fond of bee imagery from their earliest days. Their newspaper, the *Deseret News* (October 11, 1881), described the symbol of the beehive as "a significant representation of the industry, harmony, order and frugality of the people, and of the sweet results of their toil, union and intelligent cooperation." A straw skep hive became the seal of the state of Utah to represent this idea of industry.

Buddha had similar ideas centuries before the Mormons. He is recorded as saying that "a lay disciple should earn his or her livelihood the way a bee collects honey, by diligent hard work."

Bees were symbols of the desired work ethic for the earliest European settlers, for whom commercial interests and religious practice were intertwined. Honeybees were also appreciated for the stability and order in the hive, qualities sought by emigrants who had left unstable or unwelcoming European communities. In her book *Bees in America* Tammy Horn notes: "Bees fit in well in a place where Christian values merged with the Puritan work ethic to create a culture in which industry and efficiency would equal financial reward and respect."

Honeybees often get credited with human virtues, which would be a considerable surprise to the bees if they could think of themselves as having virtues. Pope Pius XII noted in a 1948 speech, "What lessons do not bees give to men? Bees are models of social life and activity, in which each class has its duty to perform and performs it without envy, without rivalry, in the order and position assigned to each, with care and love . . . Ah, if men could and would listen to the lesson of the bees: if each one knew how to do his daily duty with order and love." He went on to attribute other virtues to honeybees, including charity, purity, peace, and respect.

Many religions have gone beyond virtue to consider bees as spiritual metaphors. Muslims say that true believers are like bees that have chosen the fairest flowers to visit for nectar, and they consider bees to be angels. The ancient Egyptians believed that bees were born from the tear of the sun god Ra falling to Earth, and the Greek philosopher Plato thought bees to be the souls of the righteous who were reincarnated. Catholic Christians consider the musical ability and eloquent sermons of Saint Ambrose to come from bees having touched his lips while he was in the cradle.

Bees also are associated with oratory, probably because their many buzzing sounds remind listeners of speech. In Hebrew, the root word for "honeybee" is "dvar," referring to speech and words. The Buddha said that the person whose speech is "blameless, pleasant, easy on the ear, agreeable, going to the heart, urbane, pleasing, and liked by everybody" can rightly be called "honey-tongued."

Honey is frequently used as a metaphor, connected to themes of a sweet life and health. The Greeks and Romans considered it ambrosia for the gods, and offerings of honey would keep the supplicant on the good side of the spirits. The Egyptians used honey as part of their embalming process and buried dignitaries with jars of honey to take with them to the next world.

The Old and New Testaments contain more than sixty references to honey, many alluding to how sweet the Promised Land will be, like honey. The idea of a land of milk and honey was also important imagery in the settlement of North America, thought by settlers to be the new Promised Land, which justified their divine destiny to push out the native tribes from their traditional territories.

Similarly, one Islamic scholar wrote, "Honey is a remedy for every illness and the Qur'an is a remedy for all illness of the mind, therefore I recommend to you both remedies, the Qur'an and honey." The Buddha considered honey to be one of the five essential medicines but advised moderation: "Like bees gathering honey, they take what they need, but they don't consume the whole flower."

All of these examples and stories have one thing in common: projection, where we connect humanlike qualities, values, and motivations to bee behaviors that have no real humanness associated with them. Even though our lives are far removed from those of a bee colony, attributing human attributes to bees does serve as a useful reflector. What we project onto bees provides a lens through which to view ourselves.

Honeybees excite mystical experiences and lesson-bearing homilies, the former requiring faith in the occult and the latter an ability to trust metaphor to illuminate human traits.

They also intrigue us to imagine similarities between bees and humankind, which leads us to create lessons from the bee world that inform our own.

There's another area where our fascination about bees infuses them with transcendent qualities, although its practitioners would argue that it's based in science: apitherapy.

o o o

Apitherapy, the medicinal use of bee products, arouses strong passions. Proponents of the therapeutic properties of honey, pollen, royal jelly, and stings consider honeybees to have curative qualities supported by testimonials rather than hard data. It's also the territory of true believers and a perilous domain in which to raise the lack of hard science.

The medical profession has taken claims for bee products with more than a grain of salt. An editorial in *Archives of Internal Medicine* assigned honey to the category of "worthless but harmless substances . . . The proponents share the characteristic of possessing infinitely more enthusiasm than confirmatory scientific data."

The main lobbying group and advocate in North America is the American Apitherapy Society (AAS), which publicly applies caution in their claims for bee products: "Apitherapy is not an approved form of treatment in the United States. The American Apitherapy Society both educates about and promotes Apitherapy, but it may not certify anyone as an apitherapist."

Still, its website provides a bewildering array of diseases and dysfunctions for which the society and others recommend products of the hive. Pollen, for example, is suggested as a treatment for everything from varicose veins to high cholesterol. Royal jelly, the food produced by worker bees to feed developing queen larvae, is even more robust, believed to help with anxiety, arteriosclerosis, and arthritis, and that's just the *A*'s.

The properties attributed to these and other bee products are vague, such as anti-inflammatory, anticancer, and antiarthritic, as well as claims that they are high in vitamins and minerals. Their efficacy is most often proven by anecdote, with first-person accounts rather than data.

Almost no systematic medical studies have investigated any products of the hive, with two notable exceptions. One is the work of Peter Molan, a New Zealand scientist considered to be the world's most recognized expert on therapeutic uses of honey. He has maintained credibility by staying outside the cultish apitherapy community, which is prone to making claims about bee products with a low bar for proof.

Honey has been on the front lines of tension between naturopaths and scientific researchers, the former driven by folk wisdom and experience passed on by generations of healers, and the latter by data-based standards for proof. Apitherapists recommend honey as a treatment for a diversity of ailments, including anorexia, athlete's foot, cataracts, conjunctivitis, constipation, eczema, insomnia, laryngitis, lip sores, osteoporosis, stomach and intestinal ulcers, and wounds resulting from accidents, surgery, bed sores, or burns.

Molan's research has elevated folk medicine around honey into rigorous science and in the process stimulated a large and lucrative industry, currently valued at about US$80 million annually, applying manuka honey for medicinal purposes. His professional signature is to challenge myth with experiment, and his more than 150 publications spanning almost three decades at the University of Waikato represent the most thorough body of research on honey as a medicinal product ever conducted.

Molan grew up in Wales and couldn't decide between biology and chemistry, so he split the difference and earned a PhD in biochemistry at Cardiff University. Molan spent four years at a Liverpool university studying how saliva affects bacterial growth but got tired of living in old gray cities and studying spit. He immigrated to New Zealand, where he found honey to be more to his liking than saliva.

Mānuka is Māori for a bushy, honey-producing shrub native to New Zealand and southeastern Australia, also commonly called "tea tree" because Captain Cook supposedly made a tea from its leaves. The dark, strongly flavored honey made from its nectar has provided a significant bloom for beekeepers since honeybees were introduced in 1839 by European settlers but more recently has become renowned for its medicinal uses.

Honey has been of interest to healers for millennia. Molan notes that "4,000 years ago the Egyptians wrote about mixing honey on cotton fibers and applying it for wound dressing materials. They also used honey in their eyes to treat diseases." Greek pharmacologist Dioscorides and philosopher Aristotle both commented that some honeys were better than others; a pale yellow honey from Attica was particularly active in healing wounds.

In New Zealand the manuka tree had a folk reputation with the Māori as an antiseptic, and that reputation was passed on to honey when honeybees arrived. At the suggestion of a local beekeeper, Molan began studying its properties more systematically. In initial laboratory tests he was surprised at how well some but not all batches of manuka honey prevented bacterial growth.

Molan first turned his attention to stomach ulcers as part of a growing wave of scientists and physicians who were realizing that it was not stress but bacteria, specifically *Helicobacter pylori*, that causes gastric bleeding. In laboratory tests, active batches of manuka honey were as good as commonly used antibiotics—and better than any other honey—in preventing *Helicobacter* growth.

An important finding from this early work was that not all manuka honey is equally effective. Eventually the variable compound was identified as methylglyoxal (MGO); as a result, each batch of certified manuka honey is tested before bottling so that only high-activity manuka honey is marketed for its therapeutic benefits.

The next step was clinical trials, but the results were ambiguous, so Molan is cautious about recommending honey for treatment of stomach ulcers. Companies selling manuka honey are less reserved, focusing on the honey's antibacterial properties without making direct claims about antiulcer activity. The web is full of recommendations from naturopaths to take three to four spoonfuls of manuka daily if you're suffering from gastric ulcers, with no evidence to date that it's clinically effective.

Manuka research then examined honey as a wound and burn dressing, and here the clinical data have been positive and the claims appropriately robust. Bacterial resistance to most antibiotics has become epidemic, and honey has been a particularly useful addition to the medical repertoire because its physical and antibacterial properties provide multifaceted modes of action effective against even highly resistant superbug bacteria. Thirty-two studies on more than 2,250 patients have created a strong database demonstrating the efficacy of honey in treating wounds and burns.

Slapping honey on a wound may not be an obvious way to heal it, but honey's thickness, stickiness, low moisture content, high sugar concentration, and antibacterial compounds have numerous advantages over commonly used dressings and antibiotics. High-MGO manuka honey has tested the strongest, and when applied generously and evenly to a clean dressing pad, it prevents bacterial growth while providing a protective barrier and maintaining a moist environment that accelerates new cell growth. Seven companies globally, from New Zealand, the United States, the United Kingdom, and the

Netherlands, now sell twenty distinct sterile-pad products with honey as the wound or burn dressing.

Even though Molan has little time for folk claims about manuka honey that have not been rigorously tested, he remains annoyed at the hesitancy of mainstream physicians to adopt practices that have considerable science behind them. In a recent article he wrote: "For evidence-based medicine to be practised in wound care, it is necessary to compare the evidence that does exist, rather than be influenced by advertising and other forms of sales promotion. Honey, the oldest wound dressing material known to medicine, can give positive results where the most modern products are failing."

But hesitant mainstream physicians aren't the only problem with manuka honey; apparently much of the honey labeled "manuka" isn't. A 2013 British study reported in the *New Zealand Herald* noted that eighteen hundred metric tons of manuka honey are sold annually in Great Britain alone, but New Zealand produces only seventeen hundred metric tons per year. Further tests indicated that, although most honeys officially accredited as manuka in New Zealand were indeed manuka, more than half of the nonaccredited honeys tested out as something else.

Authorities are investigating.

o o o

The extent of research and positive results around honey as a wound dressing are the exception; few aspects of apitherapy have been examined with similarly rigorous research, and of those even fewer have been supported by data. One other that has been carefully studied is bee venom therapy, the unusual practice of treating patients with bee stings.

Applying bee stings is dangerous since a single bee sting can be fatal if the patient is allergic. Still, sufferers of arthritis and multiple sclerosis (MS) who have not found relief elsewhere frequently find their way to a local beekeeper and submit to many stings a week in hope of relief.

Multiple sclerosis is particularly intriguing because firstperson accounts sound miraculous while scientific studies have found zero evidence that bee venom is effective. Typically patients desperate for relief make their way to a local beekeeper, who collects honeybees in a jar and then applies stings by pressing up to twenty bees into them two to three times a week, often at acupuncture points. When successful, treatments are reduced to maintenance stings, a few every week or so.

The AAS website is replete with testimonials, and it's hard to argue with its passion for bee venom therapy. A registered nurse wrote:

"Nine years ago I had to go on disability because I needed a cane and could walk only short distances, I could no longer drive, I had tremendous fatigue and intractable vertigo that no prescribed medication could help. This lasted until 2003, when I found Pat Wagner's website—www.olg.com/beelady—and ordered her book, *How Well Are You Willing to Bee?* I was able to get bees for free from a local beekeeper, who is a wonderful person, as most beekeepers tend to be. The treatment lasted six weeks, and my husband graciously gave me eight to ten stings three times a week. I am now walking independently. I'm driving without hand controls or any other type of assistance. My fatigue has eased significantly."

Another testimonial came from a Yarmouth, Maine, resident, referring to treatment by a psychiatrist, who also is vice president of the AAS: "Theo began stinging my feet and spine over a period of two years. The story of apitherapy and the bees has been my manifesto. I sing it across backyards and in coffee shops. The journey has been wondrous. I am healthier than ever. I have hope, energy, and curiosity about the future and am ready to spread the word of the healing power of the hive."

Science disagrees; there still isn't a single study that indicates improvement in MS symptoms through bee venom therapy. Nor is there a known mechanism of action similar to Molan's detailed understanding of how honey's properties are effective on wounds. Proponents point to bee venom stimulating the human immune system, which is satisfying to believers but lacks the precision scientists seek in understanding why a treatment works.

Scientific writing is much terser than first-person accounts, lacking the storytelling power of anecdote and relying on statistics to draw conclusions. In 2005 a study published in the journal *Neurology* stated, "Treatment with bee venom in patients with relapsing multiple sclerosis did not reduce disease activity, disability, or fatigue and did not improve quality of life."

Another 2005 study published in *Allergy and Asthma Proceedings* reached the same conclusion but went even further in assessing risk: "Patients may be subjected to real risks of serious allergic reactions as well as emotional and economic costs. There was little evidence to support the use of honeybee venom in the treatment of MS." The Multiple Sclerosis International Federation looked at all of the studies and weighed in with: "There is no evidence that it produces therapeutic effects."

The contrast between the passionately persuasive stories from MS sufferers and the skeptical voice of scientific reason is stark and challenges the most fundamental basis of how each of us finds truth. Our conundrum is that we are built to respond to both stories and quantitative evidence. If anything, we are influenced more viscerally by intuition and a good tale than we are by a mountain of data and rigorous statistical analysis.

If patients believe bee venom or honey calms their symptoms, perhaps they do; the placebo effect is well known in medicine. If scientific studies say apitherapy is not effective, perhaps there haven't been enough studies; as any scientist will tell you, it's awfully difficult to prove a negative.

But perhaps practitioners of apitherapy are making a jump from our fascination with honeybees into their curative powers. We humans can believe that prayer heals, a belief emerging from faith in a higher power, or that a tiny sip of the waters from the Ganges River is therapeutic because that magnificent river has been elevated to sacred status through Hindu spiritual teachings.

Apitherapy and spiritual ideas about bees have this in common: they both turn our awe about bees into magic.

o   o   o

What's clear from the range of qualities that we read into honeybees is that they are real to believers, whether followers of shamanistic practices, purveyors of religious and ethical metaphors, believers in apitherapy, or data-hungry scientists reliant on numbers for insight.

Perhaps our response to bees says more about the human mind's ability to imagine, to create story, and to devise metaphor through other species than it does about bees themselves. We have a long history of imbuing other species with human-like qualities and using them as surrogates to explore our own issues and uncertainties.

The cave dwellers in prehistoric France are thought to have painted cave walls with their spirit totem animals to build an ongoing relationship with the animals they depended on for food and skins. Aboriginal cultures in Australia used Dreamtime for visits with kangaroos, sharks, honey ants, and many other species that provided information ensuring the continuity of their way of life on the land. Pacific Northwest coastal First Nations hunters each adopt a clan animal and ask their forgiveness when hunting to ensure that game will remain plentiful.

These and other indigenous peoples routinely use animals for soothsaying to predict the future and build relationships with animal spirits that provide a vehicle through which an unpredictable future can be divined. We have lost some of our confidence in listening to nature as modern life has come to rely more on data than myth.

One value of honeybees to us is that they serve as a vehicle to explore the most unanswerable of questions, whether we prefer fact or spirit as our guide. What is that something about bees that attracts us to them as a chaperone to the deep mysteries and similarly to use the mostly unproven products of the hive to promote health and treat illness?

For one thing, we're drawn to their sociality and imagine their imperatives are similar to ours. They are collectively like us in the complexity of their collaboration and their focus on accomplishing tasks together. We read virtues into how bees seem to work hard to conduct their colony business, although their work life manifests survival characteristics forged in the crucible of evolution that facilitate the integration of many individuals into a functioning colony rather than virtue.

We also can relate to how honeybees move through and interact with the environment around them. We, too, range widely through our habitats, gathering information that we share with our family, friends, colleagues, and tribe, making communal decisions about how to interact with our environment and exploit the resources around us. If we relate to the collective intelligence of the hive on a mystical level, it's because we intuit our own combined intelligence in a way that we cannot explain as a solely cognitive function.

Honeybee colonies are also marvels of collaboration, and bees live very much in community. Nikiah Seeds uses imagery of the hexagonal cells that make up honeycomb to articulate how individual bees connect to become a colony: "We're all individuals, but we're all working for the greater good. We're never really alone. We're each in a little cell; although we're alone in our cells, my cell touches other cells on six sides." That imperative strikes a chord, resonating with our species, which also thrives when individuals are well integrated into the communal.

Another reason we relate to honeybees through stories and myth is that most of us have a personal story that includes bees as part of our own private mythologies. I've yet to mention my involvement with bees to others and not have them recall a story about their eccentric beekeeping friend, an encounter with a swarm, an incident when they were stung, or a memory that includes bees from a time when there was a family connection back to the farm.

These memories are particularly poignant now, as honeybees are threatened by colony collapse disorder. The extent of public concern is evidence that we humans feel an intimacy with bees that transcends purely economic concerns. Our affection for bees is deep, and since we often imbue the colony with human traits it's not surprising that we might imagine they feel an intimacy with us.

It's not difficult to understand how we might read spirituality into honeybee colonies or ascribe human virtues such as hard work and diligence to their daily tasks. It's more problematic to discern why there are so many believers who, based on little evidence, consider the products of the hive to be therapeutic.

One possibility, of course, is that they are beneficial, and scientific study has just not yet been applied to prove what practitioners consider obvious. Peter Molan would argue that what's apparent through folk wisdom needs to be challenged by evidence-based medicine. But costs for large-scale research trials on human subjects can be extreme, and the likelihood that profits would be considerably below research and development expenses dissuades most serious researchers from working with honeybee products. Still, the growing market for manuka honey as a wound dressing indicates that honeybee therapy can be profitable when it's genuinely effective.

Shelves in pharmacies and health food stores are groaning with natural supplements and foods that are largely untested. Our regulators expect rigorous testing for prescription pharmaceuticals but set the bar much lower when we consider a product to be natural. The low expectations for testing coupled with the high costs of medical research combine to ensure that natural bee products remain in the realm of faith rather than science.

But that still doesn't explain why a painful sting or foul-tasting pollen would have a cult following or why a bee regurgitate like royal jelly would achieve great popularity for its healthful properties. Perhaps it's a tribute to just how enthralled we are with bees that we would make the jump from fascination with colony life to belief that almost anything "bee" is good for us.

That's the fundamental difference between apitherapy and religion: apitherapy can be tested, whereas spirituality is taken on faith. But that gap between science and spirituality has not diminished the belief we have in untested remedies. Health has both components, belief and faith, and with bees, the characteristics we imagine are enough for many to ignore the lack of data and just believe.

In that way, honeybee colonies exemplify emergent properties, in which a complex system arises out of a few simple behaviors. Every behavior of bees can be explained as a simple response to their environment and each other, yet when put together into a complex colony we imagine a deeper meaning than any of the behaviors might inspire alone. That's why bees stimulate contemplation and why we trust their products to heal. What emerges for us from the totality of a bee colony is a guide with which to ponder what can't be proven. As we view their world, we use that vista as a starting point to consider our own. Whether we use the tools of science or faith, story or fact, mysticism or religion, honeybees are powerful muses that inspire our reflections about the world around us.

Bees are the bridge linking these two shores of data and belief. As Nikiah Seeds put it, bees lead us to "the places that can't be explained. The messages the bees send us are limitless if you open yourself up to it."

8

Art and Culture

A wedding dress, hockey skates, and ceramic figurines don't usually pop up in the same sentence, let alone in a book about bees. Yet these and other items have been connected to each other and to honeybees through the remarkable work of Canadian artist Aganetha Dyck, who places everyday objects into hives and encourages bees to build comb around them.

Dyck's work didn't start out that way. She is from the prairie province of Saskatchewan. In 1975, when she was thirty-eight and a homemaker with three young children, she found that she had a growing need to get out of the house. She volunteered at the local arts center, and when the director suggested she might try some art making herself, she turned to what she knew best: laundry. Dyck began washing woolen garments over and over until they shrank and had become tiny sculptures that could stand alone when placed on the floor.

A chance encounter with a beekeeper twenty-five years ago led her to combine her artistic interest in common household items with an instant fascination with bees. She described that initial meeting in an interview published in the *Mason Journal:* "My first visit to an apiary was like entering another world, a foreign land. When the beekeeper opened the lid of a beehive for me, all my senses were awakened. I became totally alive, filled with imagination. Under the hive lid is a place filled with movement, scent, warmth, sound, and ambrosia. Working with honeybees has invigorated my ability to imagine."

Honeybees will hang comb from any object put into their hives. Dyck has used that habit to collaborate with her bee partners by inserting secondhand or slightly damaged items into colonies and letting the bees construct comb.

Her most famous piece is "Glass Dress: Lady in Waiting," a comb-encrusted wedding dress and accompanying handbag, shoes, and necklace, which are now in Canada's National Gallery. More recent work includes hockey skates and small collectable ceramic figurines, as well as a piece she did in North Dakota, in which a colony of bees built comb inside a glass replica of a crumbling farmhouse modeled on abandoned local farms that had honeybee colonies nesting in their walls.

The pieces themselves are eerie, combining the machine-made and the bee-constructed. Her art not only stimulates amazement at the precision and beauty of the bees' work but also represents the dialogue we would like to have with bees if we could only find a common language.

Honeybee comb is an architectural marvel, from the wax it's built from to the combs themselves. Worker bees secrete beeswax in thin flakes from specialized glands in their abdomens, where honey is converted into the more than three hundred individual components in beeswax. The workers construct comb using their jaws and the spines on their legs to shape and manipulate the wax flakes. It takes about a half-million flakes to make up a pound of wax.

When completed, the comb consists of thousands of back-to-back hexagonal cells, precisely engineered to remarkable precisions, within .05 mm tolerance in diameter and .002 mm in cell wall thickness. Hexagons evolved as the shape of cells for a good reason; it's the optimal design that leaves no empty spaces between cells while also packing in the greatest number of cells per unit area.

Dyck appeared in a 2006 television documentary, "Bee Talker," which focused in part on exploring the interaction between art and science. In that program she described her artistic collaboration with the bees: "I began working with honeybees because they're sculptors, because they are the best architects anywhere. They're really the artists."

Her most fundamental objective in artistically cooperating with the bees is to convey the message that bees have a capacity to mend broken objects and that we need to pay attention to their work as an inspiration to repair the environmental damage we cause. "The bee is such a tiny, tiny creature. We worry about the disappearance of the elephants, but I don't know how many of us look around to find the little honeybee, and how important that is."

Dyck described what happens when she puts an object into a hive: "The honeybees follow the object's contours, close openings, open closures. They create straight lines of comb where I would never have thought possible or necessary."

The objects the bees work with result in outcomes that Dyck could not have achieved alone and the bees would not have participated in without the objects she places in their hives. She doesn't credit the bees with any intention beyond covering objects with wax, yet our interpretation of these comb-covered objects makes us more aware of their world—and ours.

She also told me what has happened to her sensibilities as an artist from working with bees: "Being with the honeybees makes my ordinary life stand still and makes time disappear. Their warmth surrounds me. I feel connected to the ancients, to wisdom, and to collaboration. Spending time in the honeybees' world opens a new world of wonder for me."

To view her art is to approach the essence of another species, a vista unlocked through Dyck's unique collaboration with her friends and colleagues, the bees.

o o o

Why art with bees? Or perhaps it's a more general question: Why do art, and if you create art, why make it about bees? Max Wyman, noted writer, critic, and commentator on art and former board member of the Canada Council of the Arts, wrote in his book *The Defiant Imagination:* "The experience of art diverts and entertains us, thrills and exalts us, provokes us to understand and to feel compassion. The stories we tell each other affirm the importance of the human, the local, the specific; they are the crackly bits that give society texture. The imagination is the means by which we probe this mysterious and unknown realm. We find ourselves measuring life in a different way."

Art about bees traverses the range from the material to the spiritual, the practical to the sublime. The earliest depictions of bees were more prosaic than imaginative, telling stories of honey hunting and beekeeping. The first known examples of artistically represented bees are found in Spanish cave and rock paintings dated around 8000 BCE. They show complex scenes of rope ladders with human figures climbing up to what appears to be a honeybee nest inside a hollow tree. Bees are flying around the entrance, and people are waiting at the bottom of the ladder, presumably for honey stolen from the nest.

Similar cave and rock paintings have been found throughout Europe and Africa. Some show combs with darker areas representing where honey was stored and lighter-colored areas for the brood and honey hunters smoking the bees, suggesting a sophisticated understanding of both bee biology and honey hunting. Similar designs are found on pottery and gourds used to collect and store the honey.

These examples of bee art tell stories about honey gathering and also provide instructions to the next generation of honey hunters. Indeed, the ladders, containers, and methods for removing honey from wild colonies are notably similar today, from Nepal to Africa, Europe to Asia. These early paintings provide a remarkable archaeological record of people and bees from our earliest recorded history.

Beekeeping with managed colonies began emerging at the time of the pharaohs, with ceramic and woven hives first depicted in Egyptian temples and tombs decorated with elaborate scenes. These show horizontal hives stacked on top of each other and beekeepers harvesting honey and storing it in large ceramic vessels. Greek and Roman illuminated manuscripts show similar scenes, as do manuscripts from the Middle Ages and carvings of beekeeping scenes common in European churches. Along with any artistic merit they might have, these works also have historical value and provide instructional information for future beekeepers.

Bee art is often instructive, with images and carvings of bees and hives meant to impart lessons of thrift, labor, diligence, and frugality and to remind the viewer of the importance of bees and their honey and wax products. These messages are conveyed symbolically through painted and carved images on everything from currency to heraldry, inn signs to paintings, coins to medals, and sculpted elements on churches, temples, fountains, and even doorknobs.

The symbolism of bees as hard workers is generally pretty straightforward, but allusions to bees can get unusual. Lucas Cranach the Elder, a sixteenth-century German painter, painted eleven works depicting Cupid stealing honeycomb from the bees, getting stung, and seeking consolation from Venus, described by legend as saying: "Art thou not like the bees, that are so small yet dealest wounds so cruel?" Art critics believe there is allegory here, with the sting of the bees representing syphilis and the paintings a reminder of the consequences of stealing virtue from young ladies.

Jewelry made more for adornment than instruction also appeared early in historical times and continues to the present day. One of the earliest pieces was a striking ornamental work from around 2000–1700 BCE in Crete, a gold pendant showing two symmetrical, elegantly formed queen bees with their stingers and heads touching. I have a more modern decorative piece on my desk, a queen bee imbedded in a translucent three-dimensional hexagon of richly hued amber made for me by a former student, also intended for viewing rather than for utility or messaging about virtue.

Wyman's depiction of art as a tool by which imagination probes the mysterious emerges more strongly in another medium, poetry, where art with bees has reached perhaps its highest level of emotive expression.

o o o

Poetry with bee images dwells in both the transcendent and the practical, both inspirational and instructive, with considerable room for personal interpretation. Poems about bees appear in writing as early as 29 BCE in the classical Roman poet Virgil's epic *Georgics,* four books of verse about farming. Virgil was himself said to have been visited as an infant by bees that hovered around his mouth imparting the gift of poetry.

Book IV offers a stunning array of themes devoted to bees. It's part instruction on keeping bees, part a hymn to nature, and part political teaching. We can almost hear the buzzing of bees in a meadow on a summer's day as he describes nectar-producing plants:

the twice-flowering rose-beds of Paestum,
how the endive delights in the streams it drinks,
and the green banks in parsley, and how the gourd, twisting
over the ground, swells its belly: nor would I be silent
   about
the late-flowering narcissi, or the curling stem of acanthus,
the pale ivy, and the myrtle that loves the shore.

Virgil also provides some practical advice about locating hives:

First look for a site and position for your apiary,
where no wind can enter (since the winds prevent them
carrying home their food) and where no sheep or butting
   kids
leap about among the flowers, or wandering cattle brush
the dew from the field, and wear away the growing grass.
. . . .
But let there be clear springs nearby, and pools green with
   moss,
and a little stream sliding through the grass,
and let a palm tree or a large wild-olive shade the
   entrance . . .

He suggests some natural treatments for diseased colonies that today's pesticide and antibiotic-intensive beekeepers might learn from:

I'd urge you to burn fragrant resin, right away,
and give them honey through reed pipes, freely calling them
and exhorting the weary insects to eat their familiar food.
It's good too to blend a taste of pounded oak-apples
with dry rose petals, or rich new wine boiled down
over a strong flame, or dried grapes from Psithian vines,
with Attic thyme and strong-smelling centaury.

Georgics was written at a chaotic time in Roman history. Julius Caesar had recently been assassinated, and Rome was burdened with factions and division. Underlying Virgil's descriptions and instructions for keeping bees are political lessons he wants us to learn from the bees about uniting under a single strong leader:

> With the leader safe all are of the same mind:
> if the leader's lost they break faith, and tear down the
>   honey
> they've made, themselves, and dissolve the latticed combs.
> The leader is the guardian of their labours: to the leader
> they do reverence . . .

Bees have adorned poetry since Virgil's time; a search of one poetry website, poetryfoundation.org, reveals 887 poems about bees. For comparison, it lists only 145 butterfly poems, but a robust 1,173 poems about birds. Perhaps the best-known bee poem is "The Lake Isle of Innisfree" by William Butler Yeats, written in 1888, with its classic line about a "bee-loud glade."

Yeats was an Irish poet who spent much of his adult life in urban London but yearned to build a simple cabin on the uninhabited island of Innisfree, located in the Irish freshwater lake Lough Gill. He aspired to live with a simplicity similar to what the American author Henry David Thoreau wrote about his stay at Walden Pond. "The Lake Isle of Innisfree" begins like this:

> I will arise and go now, and go to Innisfree,
> And a small cabin build there, of clay and wattles made;
> Nine bean-rows will I have there, a hive for the honey-bee,
> And live alone in the bee-loud glade.

Yeats's poem reflects a common use of bees in poetry and other art forms, as a stand-in for the broader idea of a serene natural world. For Yeats and many artists, bees are a tool to help us recall time and place, particularly situations where peace, solitude, and tranquility infused an experience.

Sylvia Plath took a different approach to bee imagery, more intense, tragic, and personal, with troubling overtones of her pending suicide. She had recently separated from her husband, Ted Hughes, and had taken her two young children with her when she wrote five poems centered on bee images that were eventually published posthumously in her last book, *Ariel*.

Plath came about her bee imagery honestly, as she and Hughes kept honeybees in the last few years of her life. In the final poem, "Wintering," she writes of harvesting honey, the bees overwintering in the dark hive. In an allusion to how she herself was trying to get through a difficult period, she wrote of their winter cluster: "This is the time of hanging on for the bees." The poem, as well as the conclusion of the bee series, ends with the more hopeful possibility of spring but also the challenge of survival:

Will the hive survive, will the gladiolas
Succeed in banking their fires
To enter another year?
What will they taste of, the Christmas roses?
The bees are flying. They taste the spring.

As was true for Yeats, Plath's bees also represented a time and place, but not so much a tranquil solitude as an interlude. The hoped-for tasting of spring never materialized in Plath's life; she committed suicide in February 1963, a few months after writing "Wintering." Remarkably, she and her children were living in the same London house that Yeats had lived in nearly a century earlier when he wrote "The Lake Isle of Innisfree."

o · o · o

Poetry may be the most evocative expression of feelings about bees and beekeeping. Popular culture, on the other hand, offers a cheeky and irreverent view, with pun, comedy, and fright rather than metaphor and verse as its primary tools. Bees have been a common motif in movies that range from horror to cartoon to drama, taking everything from supporting to leading roles. They generally do not appear as themselves but rather as some odd or exaggerated product of Hollywood's imagination.

Movies and television programs filled with ferocious bees have a numbing sameness about them. They all start with foreshadowing, perhaps a lone bee on a window or a few bees on flowers, followed by a child/beautiful woman/rugged guy getting a single sting. Finally there's the mass stinging sequence, where thousands of bees cover one or many victims, who soon succumb after considerable thrashing around and screaming.

Killer bees had a couple of decades in the horror flick limelight, reaching its peak with the big-budget 1978 movie *The Swarm*. It starred an amazing array of top-tier actors, including Michael Caine, Katharine Ross, Richard Widmark, Richard Chamberlain, Fred MacMurray, and Henry Fonda, who actually did keep bees. In this monster horror epic, a giant killer bee swarm invades the United States, destroying a nuclear missile site in Texas, overturning trains, attacking children, and leaving Houston in ruins. The swarm is eventually attracted to an exploding oil slick in the Gulf of Mexico by foghorns that imitate the mating call of drone bees and there is obliterated. Fortunately for beekeeping, the movie was a flop and closed within days.

I vowed at that time, fresh out of graduate school, that I would never be involved in a killer bee movie. But then the megahit television series *The X-Files,* which was filmed in Vancouver, called me in 1995 since I was the local bee expert, and I found my ideals succumbing to the glamour of Hollywood and the challenge of wrangling bees for the screen.

Bee wrangling is not as difficult as it might sound. The trick is to create fake swarms of bees with a queen and use synthetic pheromones to keep the bees calm. Bees can be made to orient almost anywhere as long as they have their queen and are kept dry, under natural light, and away from unusual scents, giving the illusion of attacking on camera while all they are doing is being attracted to the queen and pheromones.

We have completely covered stunt doubles and occasionally the actors themselves with bees, filled entire houses with bees hanging from the walls and ceiling, and had bees fill up airplane cockpits and cover car windshields. To the great surprise of the directors, camera operators, and actors, nobody ever gets stung, as honeybees in swarms are usually quite docile.

Except once, and that was on a very tense set for *The X-Files.* This television series ran for nine seasons with increasingly labyrinthine story lines. It focused on FBI agents who investigate unsolved cases involving paranormal phenomena. They gradually uncover the presence of aliens who are planning the invasion and colonization of Earth in collaboration with sinister elements from government and the military.

The show we worked on had bees in the service of aliens who were cloning human children and needed the children to farm an unknown crop to raise pollen to feed the bees, who themselves were doing something for the aliens that was never quite clear. What was clear was that the *X-Files* had an almost limitless budget, and a huge hive the size of a large living room was constructed out of fiberglass molded to look like honeycomb.

The director wanted to coat the fiberglass with Vaseline to add a dripping effect and film in the dark under red light with smoke pumped into the "hive" for added eeriness. We warned the director repeatedly that the bees would become disoriented and that disoriented bees are likely to sting.

But the shot was everything. We went ahead and filled the fake hive with bees, and sure enough a few people on set were stung. What was far worse was that we weren't able to recover many of the wet and disoriented bees, and most lost their lives in service to the television industry.

The phrase "no animals were harmed in filming" couldn't be used in good faith here, and I did learn something that I brought to every movie set afterward. The bees come first, and on the many subsequent bee-wrangling jobs we took on, we never let a director put the bees in harm's way again.

Bee movies can also be warm and fuzzy. Jerry Seinfeld's animated *Bee Movie* easily wins the Academy Award for Cheeky Bee Movie. A talking honeybee speaks with humans and stimulates a successful lawsuit that returns all honey to the bees so they no longer have to pollinate. Plants begin dying without the services of bees, and the continuation of life on Earth depends on flying pollen collected from roses in the Rose Bowl parade by airplane to New York.

The plane is hit by lightning, and Earth's last chance is for millions of bees to catch the falling plane and help it to land. The swarm arrives, the plane is guided safely to ground, the pollen distributed, and life on Earth saved. It's cute and full of stars, but just their voices this time, including Jerry Seinfeld,

Renée Zellweger, Matthew Broderick, John Goodman, Chris Rock, and Kathy Bates. The one movie I wish I'd worked on was *Ulee's Gold,* in which bees get to play real bees. It's a moving depiction of bees and beekeeping, with Peter Fonda playing beekeeper Ulee Jackson. He's as comfortable with bees on screen as his beekeeper dad, Henry, was offscreen.

Ulee has been worn down by being the only surviving member of his Vietnam War platoon and by dealing with his jailed son and drug-addicted daughter-in-law, their two daughters, whom he's raising, his wife's death, and a couple of sleazy criminal associates of his son who come in search of cash hidden from their last caper. All turns out well in the end, criminals vanquished, daughter-in-law free of drugs, son leaving prison to learn beekeeping from his dad, Ulee's spirit refreshed.

The plot is pretty typical Hollywood, but it's the beekeeping that steals the show in *Ulee's Gold,* with its Yeats-like serenity contrasting with Ulee's otherwise troubled life. Ulee specializes in tupelo honey, a gold-colored honey from the tupelo tree highly valued by consumers. Bees and tupelo honey are at Ulee's economic and spiritual center. As he puts it early in the film, "The bees and I have an understanding. I take care of them, they take care of me."

It's unusual for a movie to show genuine beekeeping scenes, and these are real as it gets. I went to opening night, the theater full of beekeepers given free tickets. We all felt transported to the bee yard, feeling those long, long days of tedious repetitive physical labor, the heft of honey-laden hive bodies, that inexpressible feeling of turning off a blacktop road and heading down a barely passable forest path that suddenly opens up into a sun-speckled bee-loud glade with its peaceful apiary.

We could smell the uncapped wax during honey extraction and taste the honey scraped right from the comb. An argument even broke out toward the end of the movie among the beekeepers in the theater about whether Peter Fonda was using the proper-sized nails to hammer his wooden frames together. It's a heated discussion that still arises whenever the subject of *Ulee's Gold* comes up at a beekeeping meeting.

Bees have even made their way into love and erotica through music, particularly blues. One bee classic is the Muddy Waters song "Honey Bee," in which he laments his lover's straying to make honey elsewhere but hopes to hear her "buzzin'" when she returns. Waters's performances of "Honey Bee" include a spine-chilling slide guitar solo that mimics a buzzing bee.

But the most explicit song in the bee erotica genre is Slim Harpo's "King Bee," later performed by the Rolling Stones, the Grateful Dead, the Doors, and Led Zeppelin, among others. Released in 1957, it doesn't leave much to the imagination, with its allusions to the singer's ability to buzz all night and of course the requisite puns around the all-night-long stinging that's going to go on.

"King Bee" received a Grammy Hall of Fame Award in 2008 as a recording of historical significance, perhaps the first time an insect has been so honored.

o   o   o

What are we to make of this range of media in which bees are depicted, from poetry to sculpture, stone carvings to comb-encrusted wedding dresses, movies to music, and the breadth of topics that encompass the historical, instructive, ornamental, soothing, transcendent, and erotic?

What emerges from the cornucopia of bee art is the lushness of thought and emotion available when we allow the full spectrum of experiences to enrich our senses. Art with bees energizes our capacity to imagine and deepens our attentiveness to the world around us, as well as our awareness of who we are within it.

I talked with Canadian poet Renée Saklikar about why artists have been fascinated with bees. Saklikar was the right artist to talk with about bees as she collects bee poems, hundreds of them, and admits to "having a fetish for bee poetry."

At six months of age she immigrated to Newfoundland from India with her parents, and her family gradually moved west, ending up in Vancouver. She first studied English literature, then law, at university and worked as a communications and policy expert within government and as a private consultant. Her husband, a prominent British Columbia politician, encouraged her to write, and she gradually shifted her focus to poetry.

Saklikar is currently working on a series of poems she calls the Canada Project, using bees for imagery linking her own history and sense of place as she moved across Canada. Her obsession with bees is apparent in her voice when she talks about them. Although we met in my office, she unconsciously slipped into the formal rhythms of a poet at a reading. She calls it her "durga" voice, a Hindu word meaning "inaccessible" that also refers to the fierce Hindu goddess Durga.

Saklikar believes bees permeate poetry because they provide a sensual experience, rich in sound, feel, and imagery. She spoke about poetry, but her comments seem applicable to all art inspired by bees: "It's about humming, it's about industry, it's about sex, it's about death, it's about dance, it's about the names. I love the feel of bee words in my mouth. It's also linked to my awareness of poetry as making. Bees are about making things, this creation aspect, storing, gathering. Bees are kind of our familiars."

Art with and about bees can be many things: informative, message bearing, comforting, frightening, mystical, inspirational, even sexy. It reveals an ambivalence we have about bees, from fear of their stinging to projection of a close, comfortable relationship with nature that we desire but don't often achieve.

Historically, much of bee art has been informative and descriptive, presented as sculpture, paintings, and engravings that depict bees and beekeeping in the context of harvesting honey from wild colonies and managing hives. Firmly rooted in time and place, it provides snapshots of the economic and cultural importance of bees since the beginning of human civilization.

Bees have also been used to convey messages about how we think we should live our lives. Examples of bees as thrifty, dependable, and hard working predominate, virtues that bee art encourages us to emulate.

Peter Fonda in *Ulee's Gold* reminds us of how bees can be a tranquil touchstone in times of trouble. He says at one point in the movie that beekeeping "sometimes makes me quiet." A contemplative mood is one of the many attractions for those of us who return over and over again to our apiaries, and art with bees often recalls those transcendent, peaceful moments.

But as we've seen, bees can also be frightening or erotic, killers or lovers. Finally, bees are poetry offering a vista of the world as seen through their hexagonal lens, inspiring imagination, and expressing the most personal of desires.

o o o

Bee art is particularly valuable for those of us in the sciences, we who dwell deeply in data and can lose sight of the imaginary. Life is best engaged in its full spectrum rather than having to choose between the emotional and intellectual sides of our nature. Science demands proof, whereas the artist works happily in ambiguity and feeling, but these different ways of being in the world too rarely coexist.

Max Wyman described the role of art for scientists as "releasing the visionary impulse, bringing an innovative dimension to problem solving. It adds judgment and wisdom to information. Engagement with art synthesizes the rational and the emotional, the imaginative and the intuitive. Fie on the romantics who dismiss science as incidental to the grand moral plan; but fie, too, on the rationalists who discount the importance of humanistic inquiry."

My own experiences with artists guided me toward bridging that divide and exploring the creativity of artists and scientists working together. I took part in a dance project, *Experiments,* in which biologists collaborated with dancers to produce a performance piece that explored our creative processes.

The project was born in 2006 out of conversations between Gail Lotenberg, founder and director of the LINK Dance Foundation, associated dancers and choreographers, and a few behavioral ecologists who study animals in the ecosystems in which they live. Our objective was to produce a dance performance that explored similarities and differences in how artists and scientists imagine, observe, reflect, hypothesize, and experiment.

My interest in *Experiments* was stimulated initially by exploring whether dance might be a useful tool to enliven how we scientists communicate science to a public audience. But the work with dancers and my fellow scientists had a much greater personal impact than I had expected. The biggest surprise was in how movement through dance can create a tangible embodiment of data, giving research papers both an emotional and a physical form.

In one section of *Experiments* I appear in a background video describing a South American experiment with bees we conducted thirty-five years ago. We took individual worker honeybees of two different races, the aggressive African killer bees and the relatively docile European bees, and put them in each other's hives. The European workers became African-like, living faster and dying younger. This finding was a significant scientific result demonstrating the importance of colony environment in determining bee behavior.

When I first saw the dancers in this elegantly choreographed section of *Experiments,* I was overcome with the oddest bodily sensation of being physically immersed in those study hives, decades ago in time and remote in location. I could again feel hundreds of bees crawling on my hands and arms as I removed frames from the hive for observation, achieving a visceral understanding of how the quiet, passive European bees responded to finding themselves in the jacked-up world of a killer bee hive.

I felt rather than thought about the results of this fascinating experiment. Dance had made the intellectual intuitive, taking data from the page and making it experiential. This was a completely unexpected outcome of our collaboration: a scientific paper had become physically embodied through the representational movements of three powerful dancers on stage.

Perhaps every scientific experiment should have a corresponding expression through the arts to deepen and broaden the insights research data can bring to our understanding of the world. Science done well can be as imaginative as art, soaring beyond description to tell a story, offering a glimpse into the mystery.

I collaborated with Aganetha Dyck for many years, beginning in 2004, an interaction depicted in the television documentary "Bee Talker." That work examined how artistic inquiry could contribute to scientific studies and how science could expand the dimensions of art.

My laboratory was focused on chemical communication between bees at the time Dyck and I worked together. Our collaboration inspired experiments about how pheromones were deposited and moved through honeycomb, revealing a dynamic relationship between honeybee pheromones and wax in which comb serves as a repository and temporary storage area for pheromones.

But it was the art we produced together that had the most impact, particularly a project in which we made wire replicas of our hands and let the bees build comb connecting the hand of the artist with the hand of the scientist. That work represented the potential unleashed when artists, beekeepers, and scientists collaborate.

We who study and keep bees lean toward the practical when conducting experiments or managing these productive creatures, focused on collecting objective data to test hypotheses or making pragmatic decisions about when to extract honey or how to best prepare our colonies for winter.

Because we routinely conduct business with bees, we often overlook the profound lessons they provide in spheres beyond data and commerce. Bees yield insights into the spiritual, religious, and philosophical realms for those who pause to view their message through art.

Dyck's work releases the voice of bees to speak with us at the junction of the tangible and the spiritual. She creates a path between the practical perspective that characterizes those who study and keep bees and the empyreal sphere inhabited by these intriguing animals. Her art honors honeybees by evoking their self-produced imagery, yielding unique and complex statements that link the labor of the bees to our human sensibilities.

This is the gift she brings us through her work. We inhabit different sectors, we humans in ours and the bees in theirs, scientists in the world of the practical and artists in the realm of the barely expressible.

To view her art is to bridge the chasm between species and perspectives, to connect the testable hypotheses of the scientist, the creative ability of the artist, and the underlying wisdom of the hive.

9

Being Social

What is most remarkable about a glimpse into a bee colony is not how busy they are but how busy they are not. Geoffrey Chaucer in the *Canterbury Tales* first used the phrase "busy as a bee" in the late 1300s, but it's misleading. Worker honeybees are active on occasion, almost frenetically so, but they routinely spend more of their lives resting than working.

We know this because dozens of scientists, of whom I was one, glued colored and numbered labels onto the backs of tens of thousands of bees in order to identify them individually. Then these marked bees were introduced into glass-walled hives and observed for innumerable hours.

Two major findings emerged over decades of careful observation. The first was that bees spend up to two-thirds of their lives doing nothing. Based on how they spend their time, "resters" rather than "workers" would be more accurate nomenclature for worker bees.

The second was that worker bees move serially through tasks during their lifetimes. They begin by cleaning cells for a few days, then feed larvae from about five to twelve days of age, then receive nectar and pollen, move on to build comb, fan the nest, and then guard the entrance. Bees finish their lives by foraging for five to ten days before wearing out and dying somewhere between twenty-five and thirty-five days of age.

Why are honeybees hyperactive at particular times and do nothing at others? Part of the answer involves a relationship between resting and the ages at which workers perform tasks. These two aspects of worker behavior came together for research at an unusual beekeeping junction, package bee production, a curious system of managing colonies in which a few pounds of worker honeybees are shaken into wire packages with a queen in the spring and shipped nationally or internationally to initiate new colonies.

This system was popular for decades in the United States and Canada, with more than half a million packages produced each spring in the southern states and shipped north. The packaged bees would be shaken into empty hives at their destinations and grow populous enough to produce abundant honey crops during the summer.

All the honey would be taken from colonies in the fall and the hives killed off with cyanide. Harsh as it sounds, this system was more lucrative than leaving sufficient honey in colonies for them to overwinter. A combination of quarantines due to pests, rising production and shipping costs, and the development of indoor overwintering methods in cold climates more or less eliminated this macabre method of beekeeping, although packages are still produced to repopulate colonies that die naturally over the winter.

My students and I became interested in studying the colonies from which the packages were shaken. These hives show remarkable resilience in recovering from the loss of up to six pounds of bees, half to two-thirds of their population. These package-producing colonies not only survive but by the end of the summer rebound to being as populous as colonies with no packages removed and producing just as much honey. Significant worker loss occurs naturally in colonies due to predation, swarming, nest damage, and disease, so the capacity to recover is adaptive in feral situations as well as being useful for beekeepers.

We performed a series of experiments to investigate this remarkable bounce-back achievement, examining how task performance and lifespan might interact to explain how colonies recover from stress. The bees' solution is elegant. When a colony's population suddenly drops, the remaining workers begin foraging at younger ages. They also work harder by returning to their nest with heavier nectar loads than bees from more populous colonies, often carrying maximal weight to the point that they can barely take off from their last visited flower.

There's a tradeoff, though; these early foragers die younger, apparently due to overwork. The net result is that weakened colonies catch up in every measurable way by the end of the season, with adult populations as large as in unstressed colonies and with as much honey stored. This tradeoff between working hard and dying young is highly adaptive for colonies but tough on individuals, who sacrifice their personal longevity for the good of the hive.

These observations connect resting and the ages at which bees perform tasks. Bees in unstressed colonies are restaholics rather than workaholics, but when required they can ramp up by compressing the normal time frame of work into a shorter, more intense period. Inactive bees increase their activity when stressed and perform tasks at younger ages. Colony productivity accelerates, but many of the adult workers literally work themselves into an early grave.

Bee colonies retain an unused potential for work that can be drawn on when colonies are stressed. The term "resting" implies that it's a neutral function, but it actually has significant survival value by providing a reserve force for the inevitable, unpredictable challenges facing societies.

There's an obvious lesson for stressed humans here, which is that rest may have an important relationship to our lifespan and provide resilience to respond to challenges in our personal, professional, and community lives.

Insights into resting and labor in bees also provide a window into the great dilemma for social organisms, how to organize complex societies and allocate personal workloads among many tens of thousands or millions or even billions of individuals. Understanding how honeybees organize work and make decisions can provide some useful awareness about how we can best function individually and collectively in our human organizations and societies.

It's an essential question for us as we struggle with dispersed work environments, expectations to multitask between our personal and work selves, frustration with politics and governance, and new technologies that require considerable shifts in how we access, share, and act on information.

o o o

The key to how bees work is that they do many things during a lifetime, but they do them one at a time. Bees work serially rather than by multitasking, an aspect of labor that provides an effective combination of specialized task performance and the flexibility to allocate the workforce where it's needed.

Bees accomplish this unusual blend by moving through a series of tasks as they age, learning and performing each one effectively before moving on to their next responsibility. They make radical career changes every few days, going through seven to ten or more "professions" during their brief lives. Expressed in human time, the jobs they accomplish would each require years of retraining and on-the-job experience before reaching a level of competence. Yet, bees learn their tasks and perform them effectively within hours of embarking on new work.

Worker honeybees start their careers as janitors when only a few hours old, removing cocoons and excreta from used cells in the comb. They also eliminate debris from the nest, including moldy pollen, old wax cell cappings, and dead brood or adults. Their first major job switch happens a few days later, when they become cooks and nurses. These jobs involve producing food for the larvae in specialized glands and then feeding it to them mouth to mouth.

Food processing is the next major job change, around ten to fifteen days of age, when workers receive nectar from incoming foragers and deposit it into cells or pack down pollen left in cells by returning foragers. Next up is construction, secreting wax and building the precisely ordered cells that make up comb.

At about twenty days of age our now very experienced bees join the army, standing at the nest entrance and guarding against intruders. Finally, once their few days of military service are over they begin their last career, as farmworkers harvesting crops. The foraging workers quickly learn to orient themselves to their outdoor surroundings, find their way to and from flowers, and become efficient at gathering nectar and pollen.

The number of tasks honeybees accomplish is notable, but the serial nature of these tasks is also worth paying attention to. Bees do one thing well for a period in their lives and then move on to the next chore, accomplishing jobs more efficiently through serial rather than simultaneous work.

The term "multitasking" had its origins in the computing world, where it defined the ability of microprocessors to do more than one task at a time. It works well for microchips but not so well for people since switching contexts leads to inadequate focus, which results in human error.

Research conclusively demonstrates that each task suffers when we try to do more than one thing at a time, our brains overwhelmed with information and choices. One classic experiment illuminated our inability to pay adequate attention to simultaneous tasks.

Psychologists Christopher Chabris and Daniel Simons showed volunteers a video of six teenagers weaving in and out among themselves in front of a bank of elevators while passing a basketball. Three of them are wearing white T-shirts and three black. Viewers were instructed to count the number of times the players wearing white passed the basketball. But there's a twist: midway through the video, an actor in a gorilla costume slowly walks through the group of teenagers, stopping to beat his chest. The real experiment was to see how many of the viewers observed the gorilla; most miss it completely, suggesting that we're not able to concentrate on more than one thing at a time.

Information barrage from multiple electronic sources is particularly distracting. Almost every study has shown that even youth are less effective at doing tasks when also paying attention to e-mail, texts, Facebook, Twitter, video, or audio.

The consequences of overestimating our ability to multitask are significant. Accidents are four times more likely when driving while talking on a cell phone, equivalent to driving while drunk, with twenty-six hundred deaths and 330,000 injuries annually in the United States caused by distracted drivers using cell phones. Aviation hasn't been immune; four people were killed in a spectacular 2011 crash of a medical helicopter in Missouri while the pilot was texting while trying to land.

Why do we insist that we can multitask in the face of overwhelming evidence that task performance suffers when we try to do more than one thing at a time? Elizabeth Cohen, a cognitive science professor at West Virginia University, has achieved Internet notoriety as a quirky professor who casually presents her online course lectures on multitasking from home in her pajamas.

Pajamas or not, she's effective in explaining that it's primarily cultural pressure that implies we can multitask: "We idealize people we see juggling multiple things, oh wow, that person can handle it all. You'll see job ads that specifically request people who can multitask. We're overconfident in our ability to handle and benefit from multitasking, it makes us feel good to think we're doing so many things at once."

Of course our lives are more complex than a bee's, but we are similarly challenged with many tasks that need to be done each day in our families, jobs, and communities. How can we best perform that work? The lesson from bees is to do tasks in order, not simultaneously.

It's about focus. Bees may not be intelligent, but they are single minded. There is a presence to bees as they go about working that suggests complete absorption in the one task that they are performing at that point in their life.

Focus expresses an ability to concentrate. Although bees are unlikely to have any cognitive understanding that they are focused, their behavior exhibits those aspects of concentration that result in highly efficient and effective work. Our human term would be "present in the moment," attentive to the one task they are performing to the exclusion of any distractions.

An observer comparing honeybee and human behavior would notice immediately that we appear distracted and disengaged relative to bees. Our contemporary human working style is to attempt many things at once, doing none of them well, while bees do one thing at a time and perform that task with great precision and effectiveness.

Can we learn to better focus if we spend time with bees? I have met and talked with thousands of beekeepers, individuals diverse in their personalities and in every cultural or economic way imaginable, but they share a calm focus that descends as they walk into their apiaries. They exhibit simplicity of thought and heightened efficiency while in the bees' world, absorbing that sense of presence manifested in the hives they open.

It's bee time, the perfect antidote to our multitasking modern life.

o   o   o

Serial tasking based on age seems like a smart way to work, combining the benefits of specialization with those of being a generalist. But colonies must also have flexibility to shift workloads if they are to respond successfully to changing internal needs and external challenges.

Jobs at a human workplace such as a construction site or an office are assigned by a manager or a supervisor. But bees don't have managers or supervisors, and individual honeybees are aware of only a small region in their colonies, raising the question of how each bee decides what to do without any colony-level command.

Answering that simple question has required lifetimes of complex experiments and the slow, steady, patient revealing of information that is the bread and butter of scientific investigation. One such lifetime has been that of Rob Page, whose forty years of research on how bees regulate pollen collection has revealed a few simple rules that result in complex but well-coordinated integration of work.

Page is an American success story, born dirt poor in a farm labor camp in California, the son of Dust Bowl refugees from Oklahoma. He lived there until his father got a regular job driving a truck for a bread company, and the family moved to Bakersfield. He's come a long way since then; undergraduate studies at San Jose State University led to graduate work in entomology at the University of California at Davis, where he eventually became a professor and chair of the same department he had been a student in. Currently Page is university provost and foundation chair of Life Sciences at Arizona State University. Unusual for a high-level administrator, he maintains an active research laboratory.

The work carried out by Page and his many colleagues has revealed that work allocation in honeybee colonies results from tens of thousands of individual choices based on each bee's immediate environment. His 2013 book, *The Spirit of the Hive,* summarizes the findings that have emerged from his more than two hundred research publications dissecting bees' foraging decisions, examined at many levels of biological organization from genes to behavior.

I talked with Page between his return from a research trip to Panama and his departure for a trip to Germany, where he has been collaborating with European scientists. He told me the most surprising discovery in his research career "was the one that should have been obvious from the beginning. It was . . . literally one morning when I was taking a shower [that] it hit me: A worker honeybee is just an insect. The simplest models are best. One cannot observe a hive of honeybees without getting the feeling that they are engaged in highly coordinated and cooperative behavior. The coordinated behavior long observed and admired emerges from a simple logic of self-organization and requires only that worker honeybees respond to stimuli that they encounter."

The two simple stimuli Page has studied most closely are messages from larvae that tell workers they need to be fed and messages from stored pollen that inform bees how much pollen is available for feeding larvae. These stimuli are in a dynamic relationship with each other and must be well balanced for efficient brood rearing.

The larvae inform worker honeybees that they need food primarily through a multicomponent pheromone they secrete. Adult honeybee workers produce food in glands that metabolize protein-rich pollen into easily digested brood food, which is then regurgitated and deposited in cells containing the hungry larvae. The larval pheromone stimulates worker bees to activate their glands and seek pollen stored in cells to consume and transform into larval food.

Pollen collection is regulated by how easily foragers find cells in which to deposit their pollen. When the cupboard is full, it takes a long time for foragers to find empty cells and unload their pollen, and so they reduce their own foraging as well as their recruitment of other workers to forage for pollen. When cells are empty and demand is higher, motivation to continue their own pollen foraging is high, and the foragers dance vigorously, their way of recruiting others to join the pollen-collecting brigade.

It's a well-balanced system with both short- and long-term components. In the short term, hungry larvae stimulate nurse bees to retrieve pollen, creating empty cells that inform foragers that more pollen is needed. When there are few larvae in the nest, the pollen accumulates, and foragers are diverted from pollen and concentrate more on nectar.

The longer-term component involves flexibility to shift the ages at which workers perform particular tasks. For example, if colonies consistently have many larvae that need care, adult workers will start their nursing tasks at younger ages and continue as nurse bees for a few extra days. Or if colonies are short of pollen, some bees will start foraging for pollen at younger ages. Hormones that regulate transitions between tasks mediate the bees' age-specific shifts in job performance. These hormones are programmed to increase over a bee's life, with new chores beginning as higher levels of hormone are reached. But bees are constantly gathering input on colony conditions. Stimuli such as larval pheromones, the availability of empty cells, inclement or favorable weather, season, nutritional status, and the age structure of colonies can retard or accelerate the pace at which hormones increase, thereby slowing down or speeding up the ages at which jobs are performed.

There is another factor that adds diversity to the colony's capacity to respond to changes in what work is required, genetics. Not every bee is the same; some are more likely to collect pollen than others, based on having lower thresholds for the factors that stimulate pollen collection. Genetic diversity turns out to be a key element for effective colony decisions as it buffers the colony from reacting too weakly or strongly to stimuli.

Thus, through this simple system the decentralized, independent decisions of thousands of individuals accumulate into the higher-level response of colony functions. This honeybee system is strikingly similar to a very human phenomenon, crowdsourcing, in which tasks are distributed over a large number of people, with each contributing a small amount to a much larger endeavor.

We've come to assume that almost every idea originated online, but the concept of crowdsourcing predated computers by a century or so. The classic example of crowdsourcing came from British scientist Francis Galton, who encountered a contest to guess the weight of an ox at a local fair in 1906. Few individuals were even close, but when Galton added up and averaged the 787 guesses, the result came out to 1,197 pounds, only one pound off from the actual weight of 1,198 pounds. The *Oxford English Dictionary* adopted "crowdsourcing" even before the idea was articulated by Galton, issuing a call for quotations in the mid-1800s that illustrated the proper use of words in the dictionary. It received six million submissions over a seventy-year period.

In his book *The Wisdom of Crowds* James Surowiecki, a business columnist for the *New Yorker,* illuminates why crowds are more knowledgeable than individuals. He describes how the collective intelligence of groups leads to the best decisions and generally provides performance superior to that of a single or a few decision makers: "We're programmed to be collectively smart. Groups are remarkably intelligent, and are often smarter than the smartest people in them. The idea of the wisdom of a crowd is not that a group will always give you the right answer but that on average it will consistently come up with a better answer than any individual can provide."

Surowiecki goes on to analyze why crowdsourcing provides better outcomes than individual decisions. Qualities contributing to the success of crowd wisdom include the independence of each decision, the diversity of decision makers so that biased individuals don't unduly influence decisions, decentralization so that outcomes draw on local knowledge, and aggregation, the capacity to turn individual judgments into collective action.

Research on pollen collection has found each of those qualities in honeybee colonies. Each worker bee makes decisions on what task to perform independently of other bees, diversity is ensured by the genetic variance in a colony's worker bees, decisions are decentralized and based on information from each bee's local experience, and those individual bee actions aggregate into effective colony work performance.

Can bees help us to find the wisdom in crowds as we ponder how to best organize our own tasks? I asked Rob Page what he had learned from his forays into bee time: "The idea that global behavior can arise from fairly simple decision-making rules left a permanent imprint on my brain. I'm frustratingly modular in the way that I approach problems and questions, but in working in complex social systems I had to work that way. To break problems into parts and then try to solve each part, then link them together—it changed me as a person, became an embedded part of my personality."

o   o   o

Colonies seem to be able to organize their workloads cohesively and effectively through a casual process akin to crowdsourcing, but in one context thousands of honeybees need to come to a more formal agreement—and quickly.

Honeybees reproduce by swarming, in which the old queen and a majority of the workers leave the hive and form a cluster on a nearby tree branch or other overhang. The swarm sends out scouts to seek a new nest site, which the swarm must agree on and move to within a few days if it is to survive to initiate the new colony.

No one knows more about how swarms make this crucial decision than Tom Seeley, a Cornell University professor who has dedicated a long and illustrious career to investigating how swarms decide where to colonize. Seeley is tall and rangy, spare in his movements, precise in his conversation, patient by nature. He has lived his entire life in New England and carries his characteristic New England reserve and wry wit well.

Much of his research has taken place on an isolated island seven miles off the coast of Maine, where there are no feral or managed honeybees beyond the swarms that Seeley brings to the island to study each summer. It's an unusual situation, as there are no natural nest sites or managed honeybee colonies, so Seeley and his colleagues can control the nest options the swarms can choose from as well as mark almost every bee in a swarm individually to examine their behaviors.

Seeley is in a department at Cornell that stresses neurobiology, which fits how he looks at swarms. In a 2012 article in *Smithsonian* magazine he noted: "I think of a swarm as an exposed brain that hangs quietly from a tree branch." His first decade or so of research focused on the characteristics that swarms prefer in choosing their nests.

It's the most crucial choice feral honeybees make because a nest site that's too large or small, drafty or wet could be fatal. The size needs to be large enough to build sufficient comb to rear brood and store sixty to eighty pounds of honey over the winter but not so large as to require too much energy to heat in the winter and cool in the summer. A forty-liter volume is ideal, about the size of the standard hive box that beekeepers use. A suitable nest also needs to be dry or at least easily caulked by bees with wax and plant resins and have a fairly small entrance that can be defended from predators. Entrances at the bottom of the nest and south facing are preferred to aid with thermoregulation.

Swarms send out a few hundred scouts whose mission is to find and evaluate nest sites. When a potential cavity is found, they spend about fifteen to sixty minutes inspecting, alternating brief flights outside the nest with rapid walking inside the cavity and occasional short inside flights of a second or so. Mapping these internal inspections yielded a remarkable result: bees use a process akin to integral calculus to process their inspection data into an estimate of the cavity size.

Scouts that have found a possible site return to the swarm and perform a dance on the swarm face that is shadowed by other potential scouts, a dance that communicates the distance and direction to the nest. These followers then depart and inspect the promising location, and, if suitably impressed, they too will return and dance. The dance has a figure-eight pattern, with a straight run during which the dancing bee waggles her abdomen and buzzes. The length of the buzz communicates distance to the nest site, while the angle of the straight run relative to the sun tells the followers its direction.

Here's where Seeley's work really gets interesting. Scouts discover about twenty-five potential nest sites on average, and somehow the swarm debates the sites' qualities, comes to agreement on one, and then lifts off and flies cross-country to inhabit its new home. Group solidarity is essential. There's only one queen, and workers without their queen will die within a few days, so there's no option for a split decision; the swarm must stick together.

Nests that seem superior receive more vigorous dances than those that are mediocre or average. Both the rate of the dance and its duration are greater for better nest sites, thereby communicating the scout's evaluation of its quality. New scouts are more likely to follow and respond to a vigorous dance, and so the numbers of scouts and dancers to the best sites grow over a few hours or a day or two.

But what happens to the scouts dancing to the inferior sites? These scouts perform lazier dances and recruit fewer new scouts and gradually reduce their own visits and stop dancing. Without new recruitment those sites soon fall out of consideration. When only one site is left, the swarm takes to the air and flies to the new nest, where it quickly begins building comb and initiating foraging and brood rearing. Apparently its decision to fly isn't based on consensus because that would require ten or fifteen thousand bees to agree, an unlikely scenario.

Rather, Seeley believes that flight is initiated when a quorum is reached, a minimum number necessary to make a decision. In swarms, Seeley's studies have revealed that seventy-five to one hundred scouts are sufficient to trigger swarm movement. These bees produce a distinctive piping sound that stimulates bees to prepare for takeoff once a quorum of bees agrees on the same site. Occasionally two strong nest-site contenders remain in the running when swarms lift off. They split and fly in two directions for twenty or thirty yards, realize they are divided, and return to the cluster site to continue the discussions until one site predominates.

Seeley's New England background steered him toward recognizing a strong congruence between honeybee nest-site selection and how town meetings function, with both using similar processes of collective decision making. In his book *Honeybee Democracy* Seeley writes: "In both bee swarms and town meetings, the way the group selects its course of future action is by staging an open competition among the proposed alternatives. Each listener makes an independent assessment of the proposal and decides whether to reject or accept it, and those that accept it may announce their own support for the proposal. The better the proposal, the more supporters it will attract."

Seeley points out that swarms and town meetings have much in common. Both are groups that have shared interests in successful outcomes and minimize the importance of a strong leader in favor of collective wisdom. Both entertain diverse solutions to issues and then aggregate the group's knowledge through debate. Finally, both swarms and town meetings utilize quorums in decision making to facilitate cohesion, accuracy, and speed, defining ahead of time whether decisions require a simple majority, 60 percent, two-thirds of members, or some other figure such as seventy-five to one hundred bees in the case of swarms.

Seeley has absorbed the lessons from collective decision making in bees into his own life, particularly in developing a management style for his role as chair of the Department of Neurobiology and Behavior at Cornell. His many, many hours spent with bee swarms have been deeply influential and affecting: "Some have said that honeybees are messengers sent by the gods to show us how we ought to live. This story about the bees provides useful guidelines to human groups whose members share common interests and want to make good group decisions. Of course, employing insects as management gurus has its limitations. Nevertheless, I will claim that the bees demonstrate to us several principles of effective group decision making and that by implementing them we can raise the reliability of decision making by human groups."

o o o

Can the honeybee model for organizing work and making decisions really be useful in human groups? Perhaps the best way to answer that question is to start by considering where our governance is not beelike.

The most antihoneybee model I can think of is exemplified by the current situation in the US political arena. "Arena" is the right word, as politics has become more closely aligned with the octagon in ultimate fighting than the genteel decision making of a New England town hall meeting. The nature of political discourse and governance today is characterized by a toxic mix of divisive and entrenched positions on issues, depending on a few leaders exercising power rather than a more dispersed decision-making structure.

Problems in contemporary governance are widespread, local to national. Special interests and lobbyists abound, supported by experts on all sides of issues whose practiced voices and professional opinions overwhelm citizens. Government seems to lack trust in voters to understand and make decisions on complex issues. The US Congress is gridlocked, having lost the capacity to listen and compromise.

This breakdown in our democratic institutions is bemoaned daily in almost every media report about policy or politics but on the bright side has generated an epidemic of interest in participatory democracy. Perhaps no organization better represents the growing participatory movement than America-*Speaks*, founded in 1995 by Carolyn Lukensmeyer to address this democratic deficiency. Since then, the organization has brought together 170,000 citizens to deliberate dozens of critical policy and community issues.

The core of this nonprofit's method is its "21st Century Town Meeting," a version of the traditional New England town hall scaled up for thousands of participants. Lukensmeyer writes in her 2013 book, *Bringing Citizen Voices to the Table*, that she considers these meetings as "a deliberative process through which groups of citizens representative of their communities learn, express their points of view, and discover common ground."

The organization's strategies for exploring issues and generating decisions are similar to swarm behavior, in which information is provided to small groups of bees by scouts, and those nodes of information gradually expand to encompass the larger swarm and result in a decision about which nest site to inhabit. America*Speaks* utilizes its own equivalent: small-table discussions with ten or twelve participants, networked through computers at each table that connect dozens or hundreds of these breakout groups.

Up to twelve thousand citizens have taken part in a single America*Speaks* event, sometimes at one large convention center but often simultaneously at dispersed but electronically connected sites across the United States. Analysts distill each table's conversations, summarize the results on large video screens, and send summaries back to tables for further discussion and eventually for votes. The participants have individual keypads through which they can vote, ensuring independence and promoting a sense of empowered participation. A preliminary report is generated by the end of the day and sent home with each participant.

Seemingly intractable problems have yielded to this crowd-sourced wisdom, reminiscent of collective decision making by honeybees. In 2010 America*Speaks* conducted a federal budget exercise with thirty-five hundred citizens at fifty-seven sites across America, with participants from conservative Tea Party backgrounds sitting down with liberals from MoveOn. Young and old, rich and poor, all races and ethnicities were represented at the breakout tables. Each table was tasked with developing a plan to reduce the federal deficit by the target Congress had set but proved incapable of reaching: $1.2 trillion. Sixty percent of the tables succeeded, and 75 percent managed to reduce the deficit by at least $800 million through budget cuts and tax increases.

Another America*Speaks* project was the Unified New Orleans Plan, a process designed to break the deadlock after Hurricane Katrina in making decisions about how to rebuild the city. Government, with its access to abundant experts, had failed to develop a plan, with racial and economic disparities among citizens a major factor in shooting down every project government proposed. America*Speaks* was brought in to help create a cohesive plan in this very disunited and challenging situation. It brought together planners, officials, and citizens with two key elements guiding the process: the participants needed to be representative of residents and were authorized to develop a plan that government would adopt.

America*Speaks* held two public forums, involving citizens still living in New Orleans as well as those who had fled the floods and not yet returned. The refugee residents gathered at various locations around the country close to where they had temporarily relocated. Their website reported that, at the forums, "citizens discussed how to ensure safety from future flooding, empowered residents to rebuild safe and stable neighborhoods, provided incentives and housing so people could return, and established sustainable, equitable public services." By the end of the second forum, 92 percent of the four thousand participants had agreed on a comprehensive rebuilding plan that served as the basis for reconstructing New Orleans.

It's striking how similar these and other participatory processes are to what goes on in honeybee colonies and swarms. For one thing, both manifest decentralized decision making, with participants sharing information with those in their immediate vicinity. All information is fully available to each individual; there are no back-room meetings or hidden reports or agendas.

The operating unit in human participatory democracy is the breakout group, whereas in bee colonies it's the small group of bees in each other's immediate vicinity who share a particular task at that time. For swarms, it's the bees following each dance that commit to exploring the same nest site if its characteristics seem favorable.

The breakout system results in dozens or hundreds of nodes at which information is gathered, communicated, and acted on. Each individual at each table, comb site, or location in the swarm evaluates information and then gravitates to the decision or action the individual feels is most appropriate, reducing the influence of special interests and experts.

These independent decisions are aggregated by the entire colony, swarm, or human meeting, in person or online. Higher-level outcomes emerge from the accumulated set of opinions and behaviors. Thus, the processes shared by participatory democracy and honeybees build constituencies committed to agreed-upon outcomes (i.e., buy-in) that result from the gradual coalescing of opinion around the best ideas.

Another common component is that diversity is key to success. Participatory democracy works best with a broad swath of income, ethnic, age, gender, and political representation. Similarly, we've seen that honeybee colonies function optimally with worker bees representing an array of genetically based probabilities to perform particular tasks, as well as a wide range of ages available for work assignments.

Still another important and shared aspect of human and honeybee governance is that both benefit from reaching decisions collaboratively, working together for outcomes that emphasize the common good. One segment of the population isn't favored over others, but it's the big unit that's favored, the colony for bees or all residents of New Orleans rather than just the privileged.

We share with bees another facet of governance, the tragedy of chaos when events upend the peaceful order of a well-functioning society. Human history is more than replete with war and violence, considerably more so than the honeybee world, but even the placid world of the honeybee colony can descend into conflict if the queen dies suddenly. The workers sense her loss within minutes due to diminished queen odors in the hive and begin rearing a new queen within hours (see the next chapter for a full discussion). Occasionally they fail, and the beekeeper's phrase for that situation says it all: the colony becomes hopelessly queenless.

The hive will dwindle and die within a few months, but as it declines worker bees become highly aggressive toward each other and any unfortunate beekeepers who might open the nest. Some of the workers develop their ovaries and lay eggs, which develop into male drones since they are not fertilized. Vicious fights break out between workers trying to become dominant as egg layers, heartbreaking acts of desperation because, dominant or not, their demise and that of the colony are inevitable without the rearing of new female workers.

This underlying potential for conflict is all the more remarkable considering the harmony that usually reigns within the hive. It's a reminder that the most cooperative of societies, even that of honeybees, can collapse into disorder and violence given the right circumstances.

o   o   o

We've seen numerous parallels between how bees work and make decisions and corresponding human strategies that are suited to contemporary human societies. We—and bees— benefit from performing tasks serially rather than multitasking, and we both take advantage of crowdsourcing concepts in evaluating information and reaching wise decisions.

Another successful practice we hold in common is decentralizing decision making to a local level and aggregating solutions from numerous independent individuals when higherlevel resolutions are needed. Bee and human societies thrive when the diversity of decision makers is broad and when we rely on collective wisdom rather than a small circle of isolated leaders.

Many strategic-thinking gurus have reached the same conclusions. It's notable that two independently evolved social systems that are only distantly related, humans and honeybees, developed similar ideas about how to perform work and make decisions.

David Zinger, a management consultant and sought-after speaker, has implemented attributes of honeybees into his practice running the six-thousand-member Employee Engagement Network. He and his clients work to improve organizational strategies by focusing on how groups can improve their teamwork.

His bee project, named "Waggle," was itself a collaborative endeavor with artist Aganetha Dyck and a beekeeper, Phil

Veldhuis, who moonlights as a philosophy professor at the University of Manitoba. They were inspired by placing common workplace items such as computers or calculators into hives and surveying how bees worked collaboratively to surround the objects with comb. From their observations they developed a thirteen-point guide to the attributes of honeybees that can provide insights into how human organizations might improve their internal dynamics.

Among the colony metaphors Zinger uses to guide organizations are ideas such as "waggling," focusing on using information transfer in hives as a template for how data can best move through human groups, and "succession," addressing issues of flexibility in work performance by directing clients' attention to how honeybees make continual adjustments that respond to external factors. Also included in the list are "go girl," encouraging strong roles for women, and "innovate," reminding participants that the best protector against organizational collapse disorder is to build a strong, resilient, and interdependent community similar to a honeybee colony.

There are, of course, differences between honeybee societies and ours, and colony metaphors do have their limits. For one thing, we value personal fulfillment and individual achievement more than bees. Bee societies are purely an outcome of natural selection; while human societies were ultimately forged through evolution, we also include cultural and cognitive components in making decisions about how we want to structure our social systems. And there certainly are governance models differing from those of honeybees that have been successful in our work and political spheres, whereas bees show relative inflexibility in how their colonies are organized.

Still, what is unique and particularly useful about bees is the affinity we feel for them as a similarly social species. Their behavior can be a window through which to ponder our own.

Because we share that fundamental sociality, we can learn by observing their lives and comparing the behaviors that determine whether we—and they—thrive

Spending time with bees also provides opportunities for personal insights into how each of us wants to be in relation to the world around us. We who work closely with honeybees find ourselves becoming more like them, an impact not well articulated in scientific or beekeeping circles.

Rob Page, for example, gained a perspective about how he wanted to work by studying bees, first as a scientist running a laboratory and then as an administrator in charge of a vastly complicated faculty. His personal style developed directly from spending time in apiaries and observing how bees behave. From them he learned to break problems into parts, ask simple questions, and then let the results aggregate to form a complex tapestry.

Tom Seeley absorbed lessons about decision making from bees. He came to practice their habits in his professional life through years of observing swarms, coming to understand that collective wisdom was as powerful in human societies as it is for bees seeking a new home. As he puts it, "Living in groups, there's a wisdom to finding a way for members to make better decisions collectively than as individuals. One valuable lesson that we can learn from the bees is that holding an open and fair competition of ideas is a smart solution to the problem of making a decision based on a pool of information dispersed across a group of individuals."

I, too, have learned much from bees, particularly one lesson that serves as the underpinning for my own work and decision making: slow down, pay attention, focus, be present. These teachings could have come from other sources, and, for some, yoga, meditation, running marathons, or a similar influence has produced a similar outcome.

But for me it was bees; I realized that we share with bees the challenge of balancing individual behaviors with social imperatives. Bees provide a reflective sounding board through which to become more aware of how we communicate, the power of collaboration, and the importance of focus and presence in daily life.

What I experienced within their colonies preceded my adoption of a very human leadership practice: dialogue. The essence of dialogue is deep conversation, listening without judgment, a capacity I learned among the bees before taking it into the human realm.

10

Conversing

A snapshot of any moment in a honeybee colony will reveal some bees resting and others working, but one other activity stands out: discussing. Bees aren't communicating in any verbal way that's familiar to us; nevertheless, they are deeply engaged with each other, passing along information about the world outside and conditions within the colony.

They're essentially carrying on a nonstop, colony-wide conversation similar to human interactions, with some of the exchanges quick and superficial while others are lengthy and profound. A quick glimpse of bee banter reveals a rapid flurry of body parts touching and whole bodies vibrating, but if slowed down to human speed we see precise movements that provide the channels through which news, data, and perceptions flow.

The most frequent behavior when worker bees meet is rapidly stroking each other's antennae. They'll do this for a few seconds to a minute or longer while also extending their long mouthparts, licking each other's tongues, heads, and legs and exchanging food. They also rub their front legs on each other's body parts. At other times we see bees vibrating while covering other bees or standing alone on the comb. Some of the buzzing is choreographed, with a pulsating bee circling in a round pattern or repeatedly turning in a distinctive figure-eight configuration.

Layered beneath these simple behaviors is an astounding array of information and discussion crucial for colony function and survival. Much of this ongoing conversation is chemical, passing queen pheromones around the nest and in this way establishing her presence and influencing many worker functions. Worker bees also talk to each other through their own pheromones, expressing alarm and using odors to orient to and from the nest. Larval bees communicate chemically with adults as well, expressing their need for food.

Bees are acutely aware of the environment outside the hive, and there's constant chatter reporting on the external world. Vibrations and dancelike chats between adult honeybees provide the most important data stream about the environment, a discussion through which foragers communicate the location and quality of nectar and pollen sources to their nest mates.

All this stroking, licking, and buzzing entails the use of language, either a vibration-based dialect that influences foraging or the chemical language of pheromones, which mediates the workers' awareness of their queen and stimulates or inhibits various worker behaviors. These interactions are how bees converse, a relentless current of news that is the cornerstone of the apparent orderliness in the hive.

It's a dialogue among thousands, a model of communication with considerable relevance to how we humans engage and make decisions as individuals and together in our communities and societies.

o o o

A verbal translation of the worker honeybees' chemical conversation might be: "How's our queen? She's still present, in good health, and laying eggs? Excellent. Long live the queen!" It's the most important discussion in the colony because if the queen is failing or dies, the workers must begin rearing a new queen within hours or the colony will not survive her passing. New queens are reared from young worker larvae by feeding them a special nutritious brood food, royal jelly. If queen rearing doesn't commence quickly after queen loss, the worker larvae soon become too old to switch their developmental pathway toward queens, and chaos will soon reign within the hive.

Beekeepers have a wonderful term that indicates the queen is present and healthy: queenright. Her absence is obvious within fifteen to thirty minutes as a sense of disarray and increasing nervousness that permeates the nest. The colony's challenge is that only a few of the twenty- to fifty-thousand worker bees in the hive actually encounter the queen, leaving the others dependent on good communication to broadcast the queenright message. There is a premium not only in knowing that the queen is extant but also in rapidly spreading the knowledge of her demise.

My students and I investigated the mystery of how workers physically distant from the queen are aware of her status by collaborating with a fine chemist at my university, Keith Slessor. It took a fortuitous accident at the end of a frustrating research day to make our most significant breakthrough.

Keith was interdisciplinary long before it was popular, and his intense appreciation for teamwork and for his colleagues was palpable. Still, getting it right was paramount, and Keith and I often terrified our students with our uncompromising conversations probing every corner of our data.

He loved nothing more than a passionate discussion, always interested in the best answer rather than necessarily winning an argument. I heard the word "bloody" so often from Keith you'd think our lab was an abattoir, but his rigor was a gift in unraveling what turned out to be quite a knotty problem. Keith passed away in 2012, but I often recall how we shared an intense obsession with data. Each afternoon found one of us running to the other's office or laboratory as soon as that day's experimental numbers came in from our students.

We had become interested in the retinue of ten to twelve worker bees that surround the queen, licking and touching their antennae to her furiously for one to two minutes each. Our hypothesis was that they were picking up the queen's pheromones and somehow transmitting them throughout the nest, but we were frustrated by the lack of a bioassay to test our hypothesis.

We had made extracts from dead queens to use in identifying her chemical signature but had no way of determining whether worker bees responded. One day, in frustration, one of our students put a dab of extract onto a glass pipette and thrust it into a cage of bees, exclaiming, "Take that, you bloody bees!" To her surprise, they formed a retinue around the glass as if it were a queen. When I ran over to Keith's office that afternoon, he had that rare smile on his face that told me we'd found our bioassay.

About fifteen years and tens of thousands of bioassays later we had identified most of the queen's pheromones, a blend of at least nine compounds, all of which are necessary to fully attract worker bees. We'd also unraveled the system by which workers pass around the information critical to the workers' awareness of the queen's presence or absence.

The solution to how bees remain aware of their queen at a distance turned out to be a thing of beauty. The retinue bees attending the queen rub and lick off her pheromones and then move through the nest acting as messengers proclaiming her presence. They pass on this chemical conversation by licking and touching antennae with other bees, which then also move on to spread her message in a similar fashion.

This transmission chain goes on for fifteen to thirty minutes, by which time most of the queen's pheromones originally picked up by the retinue attendants have been passed to other bees. But with so much pheromone moving throughout the nest, why doesn't the colony become saturated, which would prevent the workers from knowing when something untoward had happened to their queen?

The answer turned out to be the most elegant aspect of this colony-wide dialogue: the worker bees sense the queen's pheromone on contact, then cleanse their palates of her odor by swallowing some of the pheromone and absorbing the rest across their cuticle, where it ends up in their bloodstream and is inactivated or performs an as-yet unknown function as a hormone.

Thus, a continuous stream of queen pheromone is moved through the nest by messenger bees and those they contact. If the queen dies, this transmission is interrupted, and within thirty to sixty minutes any residual pheromone circulating through the nest has been absorbed. The workers respond to its absence by initiating the rearing of a new queen. We were able to verify this hypothesis by adding synthetic pheromone to queenless colonies, which prevented the queen rearing that normally would have occurred.

The core of the conversation is communicating queenrightness, but the pheromones' presence or absence also carries other crucial information. This system explains another great mystery of honeybee life: what triggers colonies to swarm. In order for colony reproduction to succeed, the worker bees need to rear a new queen prior to the exit of the old queen with a swarm. Yet, if the colony is bathed in pheromones from the old queen, and these pheromones inhibit bees from rearing new queens, why do they rear a new one prior to swarming?

One possibility is that the old queen's pheromone production diminishes prior to swarming, but we and others have found no evidence to support this idea. Another possibility is that pheromone transmission is reduced as colonies become congested prior to swarming, thereby releasing the workers from the queen's inhibitory effect.

That turned out to be the case. We put pheromone-baited lures into colonies and also sprayed excess pheromone into colonies on a daily basis. Under that regime of increased pheromone, colonies become ridiculously crowded without rearing new queens, indicating that congestion does indeed inhibit pheromone movement through the nest and triggers swarming.

Queen pheromone does many other things in this colony-wide interactive system besides alerting bees to the presence of their queen and inhibiting the rearing of new queens. For example, it inhibits the hormone that leads to foraging. By slowing the progression of hormone secretion it delays the age at first foraging, ensuring bees don't rush through their other preforaging tasks. Queen pheromone also acts to regulate genes associated with age-dependent behaviors. By activating hundreds of nursing genes and repressing foraging genes, the queen's odors delay behavioral maturation.

The queen's pheromone also decreases defensive behavior by reducing guarding at the nest entrance and lowering stinging following a disturbance. It can act as a stimulant, increasing the tempo of comb construction. Queen pheromone also plays a key role in maintaining colony social hierarchy by preventing worker bees from developing their ovaries and laying eggs.

And the nine-compound queen pheromone is only one of the many chemical messages that mediate honeybee behaviors. Some others include the alarm pheromone, which is discharged by workers when colonies are under attack. This twenty-six-component blend recruits workers to search out predators and sting. The Nasanov pheromone, which consists of seven components, is an attractant released during swarming and at the nest entrance. Another important pheromone is the brood pheromone. Its nine components stimulate nurse bees to produce brood food. It works in tandem with the queen's pheromone to inhibit worker ovary development.

These elaborate interactions reflect a complex language based on specialized chemical signals that provide syntax deep in complexity and rich in nuance. The colony's pheromone signals are characterized by intricacy and synergy, influenced by the context in which they are deployed, and mediated by their temporal and spatial distribution.

In every respect, these signals seem equivalent to the verbal exchanges characteristic of higher vertebrates. In human society, language, with its vast oral and written vocabulary, is the commerce of communication. In honeybee society, one vocabulary is an array of chemical blends and is almost certainly more complex than the more than fifty compounds identified to date.

o   o   o

The complexity of the bees' chemical vocabulary is appropriate for the intricacy of the colony's conversations, much as human communication is inordinately more layered than we understand through the deceptively simple act of talking. We exchange words as bees exchange chemicals, with cognitive, hormonal, neurological, and physiological responses that float below our conscious radar but have enormous impact on our relationships and activities.

Much as many aspects of human conversation remain shrouded in mystery, we're only beginning to understand the complexity of the queen's chemical message and the range of worker behaviors affected. One key question is why are there so many chemicals? Research to date has only grazed the surface of this compelling question, but while all her compounds are needed for full attraction to the queen and messenger transmission, individual compounds likely are more prominent for particular impacts.

One of the queen's nine pheromone components, homovanillyl alcohol (HVA), is particularly intriguing because its structure is strikingly similar to the important brain neurotransmitter dopamine. Dopamine in humans is associated with reward-based learning. Many addictive drugs act by disrupting dopamine-dependent transmission of nervous impulses, and quite a few diseases are caused by disruption of dopamine, including schizophrenia, attention deficit hyperactivity disorder (ADHD), and Parkinson's disease.

HVA in bees reduces the level of dopamine in the brain, which in turn decreases the activity level of young bees and keeps them close to the centrally located larvae rather than wandering to the edges of the nest, where food processing and storage take place. In that way it's part of the queen's influence, ensuring that newly emerged bees perform the tasks of young bees and don't age to foraging prematurely.

Considering what a verbal species we are, similar research on how conversation affects the human brain is remarkably sparse. Perhaps the most relevant study is the one carried out by Leslie Seltzer and her colleagues at the University of Wisconsin, who thought that speech might strengthen social bonds between individuals through neuroendocrine mechanisms.

Their 2012 study demonstrates that conversation decreases salivary cortisol, an indicator of stress, and increases oxytocin, a hormone associated with social bonding, pair formation, maternal behavior, trust, and empathy. Texting didn't show the same effects, even between those with close relationships like mother and daughter, suggesting it's something about live interpersonal conversation that creates the strongest social bonds.

Thus, there's evidence from both bees and humans that communication can have a profound impact on brain neurochemistry, chemical conversations in the case of bees and verbal ones for us. But does the chemical interaction between the queen and the workers and the myriad ways that workers engage with each other as they pass pheromone around qualify as conversation?

An aggregated definition from many dictionary sources suggests that bees do indeed carry on sophisticated conversations: "An exchange of observations, opinions, information, news, issues, and ideas; give-and-take of talk or nonverbal exchange."

Honeybees fit comfortably into many of the parameters that define conversation. Their interactions are exchange oriented and nonverbal, with conversationalists passing information through the concentration of odors they transfer as well as by food exchange. The information can be considered to carry opinions or news through the intensity of their stroking and licking as well as the concentration of pheromone exchanged. Intense interactions and a strong pheromone dose signal that the queen is present, an orienting worker is close to the nest, or the colony is under serious attack. Conversely, weak interactions and a low dose indicate that perhaps the queen is failing or recently dead, orienting workers are still distant from the nest, or any predator is of minor importance.

If we think about what happens in a human conversation, bees do seem to converse. Like us, they pass information, evaluate, respond, and reevaluate as new information emerges. We both pass on nuanced, complex signals perceived on many levels, some conscious and some at a subconscious neurological or physiological level.

Most significantly we—and bees—often change our behavior based on a conversation, which is one of the hallmark characteristics of a social interaction. Bees respond to each other, which is one of the core reasons we relate so strongly to them. Perhaps the only difference between bee and human conversations is that bees don't have the range of feelings we do, as least as far as we can tell.

o   o   o

As social as we are, we humans remain oblivious to the many channels of our own perception and communication as we converse. Bees can be an important bridge to those unrecognized human talents since we can study bees without the bias or blinders that come into play with human research. We may have a broader understanding of how bees perceive the world around them and how they communicate key information to other bees than we do of our own interactions.

Bees aren't only focused on each other but also survey and report on a vast area of landscape surrounding the hive for many miles in any direction. There's constant chatter in the hive about external conditions as bees returning from the field buzz and dance to share relevant information.

James Gould, professor of ecology and evolutionary biology at Princeton University, has studied how bees perceive and remember information and communicate with their nest mates through the honeybee dance language. He has great respect for the depth, breadth, and quality of that conversation. In a Canadian Broadcasting Corporation *Ideas* radio documentary he said: "Our imaginations are insufficient to understand how they sense the world. You know, if we were talking about chimpanzees, no one would have any trouble saying this was a highly cognitive behavior, weighing one thing with another, optimizing many independent factors."

The first part of a bee's environmental conversation is not between bees but rather between bees and the inanimate world. Much of this input is perception more than conversation, utilizing many channels. Bees use landmarks, odors, Earth's magnetic field, wind speed and direction, and polarized light to orient to and from the nest, exhibiting amazing qualities of recall in storing and processing that information to travel miles away from and back to their nests.

But there is some back-and-forth chatter with the environment: a plant's flowers talk with bees in a conversation that changes both bee and flower to mutual benefit. Bees carry a positive electrical charge, while flowers tend to be negatively charged, although both charges are slight. When a bee visits a flower, it triggers the flower to change its charge to positive within seconds.

Subsequent bee visitors detect this subtle difference in electrical field and avoid the flower, which is unlikely to have a nectar or pollen reward as it was just visited by another bee. This back-and-forth between bee and flower aids both, as it keeps the bee from wasting time and energy on an impoverished flower and for the plant enhances pollination by increasing the number of flowers visited. It's a conversation, albeit an electrical rather than a verbal one.

Foragers returning from flowers with nectar or pollen are not only provisioning the nest but also recruiting other bees to exploit floral resources. Here is where the most intense conversations about the environment outside the hive take place, as foragers share information about where they went and the quality of what was there.

The idea that bees are passing along information about the location of nectar and pollen sources was first demonstrated by the philosopher and writer Maurice Maeterlinck more than a hundred years ago. He let a forager find a dish of sugar syrup and return to the nest, then caught her as she left again, preventing her from returning to the dish. Yet, many other bees quickly appeared at the dish, suggesting that some information about its location was transferred from forager to nest mates back in the nest.

At first it appeared that floral odors were the recruiting information provided by successful foragers, but recruits arrive at dishes of sugar syrup that are many miles away and downwind of the hive. The German scientist Karl von Frisch and many others eventually demonstrated that foragers perform dances that communicate the distance and direction to and the quality of resources.

The most elaborate dance is the figure-eight waggle dance, which consists of a straight run during which the returning forager vibrates, turns in one direction, and makes a semicircular turn back to the starting point, vibrates in another straight run, turns in the other direction, and repeats. The dance is followed by potential forager recruits, who learn the distance to the resource by the length of the straight vibrating run, its direction by the angle of the dance on the comb relative to the sun's position, and its quality by the intensity of the vibrations.

Beneath that simple explanation is an astounding array of subtle and highly nuanced exchanges of information. For example, dance followers ask for more detailed information from the dancer by emitting a squeaking sound lasting only a tenth of a second. That peep causes the dancer to stop and give a sample of her load to the squeaker, providing direct information about the resource's quality and odor.

The quality of the flower's nectar and pollen is related through a number of dance attributes that provide backup to the primary information in the vibrating run. The lateral extent of waggling, total number of cycles, and intensity of vibrations all add to the message, while the followers encourage a dancer carrying a superior load by their enthusiastic and rapid reception of the nectar she exchanges with them. Further, the enthusiasm of recruits is context dependent, enhanced if the colony is experiencing famine conditions, when foragers dance more intensely and foragers are recruited more easily to lower-quality resources.

The subtlety and types of vibrations are notable, providing an array of communication that we are only beginning to unravel. In addition to the waggle dance, bees do a vibration dance, in which one worker grasps another and vibrates her abdomen up and down at high speeds. This action stimulates the recipient worker to move to an area in the hive where waggle dancing takes place and in that way raises the hive's excitement level for foraging. The queen is similarly vibrated just prior to exiting with a swarm.

Honeybees also perform a tremble dance, in which a returning forager will shiver if receiving bees are not taking her nectar load fast enough. She shakes her body, rotating by about fifty degrees every shake, moving slowly across comb for thirty minutes or so. This buzz is designed to ensure that nectar processing and nectar intake are aligned.

Listening to the full array of chemical and vibratory messages inside the nest is a full-body experience for bees, all senses in motion to receive information and respond. The antennae are sampling pheromones and floral odors, the tongue and front legs tasting fresh nectar and pollen from the field, specialized organs in the legs detecting vibrations passed on directly from bees and transmitted through the comb, and hollow chambers at the base of the antennae acting as ears to hear airborne sounds.

Occasionally, and only if I'm particularly calm and focused in the apiary, I've been able to put myself into a bee point of view and feel what it's like to experience this cascade of multichanneled conversation. And it's interactive, with bees receiving information, passing it on to other bees, and responding with work appropriate to the colony's immediate and long-term needs.

Bee time. It reminds us of the many channels for communication we tune into, some consciously and others existing under our radar. It can be a full-body experience for us as well when we reach the level of presence and concentration that characterizes the deepest layers of dialogue.

o o o

I had thought the experience of immersing myself into this many-channeled colony conversation was unique to the apiary and so was startled to realize that the new world of dialogue I entered after taking on a new role at my university in 2002 felt eerily akin to the inside of a bee hive. Human dialogue, when done well, has much in common with the depth and complexity of the ongoing conversation honeybees hold within their nests.

What struck me when I began directing my university's Centre for Dialogue was how alike human and honeybee dialogues are in being composed of tapestries of small communicative moments. They both produce cumulative effects from many interactions, building one on the next, with the signature quality that listening attentively is a key element.

The interactions between bees in their colonies and we in our dialogues are similar in sampling many opinions, bouncing information and ideas between participants, and building consensus slowly but inexorably, resulting in individual actions that accumulate into a group outcome.

The first formalized use of dialogue was recorded in ancient Greece and set the template that still comes to mind when we think of dialogue. Plato wrote his early dialogues sometime shortly after the death of the classical Greek philosopher Socrates in 399 BCE. Plato's Socratic dialogues record many of the conversations between Socrates and a group of his disciples who gathered to learn through what became known as the Socratic method.

The method involves sharp conversation between two or more individuals who use their intellectual talents to challenge each other's debating points. It's a dialectical approach in which opponents square off verbally and try to persuade each other, using clever questions and answers designed to entrap participants in contradictions, weakening one point of view while supporting another. The resulting exchange can be intellectually brilliant and rhetorically stylish, and the Socratic dialogues are among the great discussions of Western thought. Unfortunately, the template set by Socrates defined the culture of dialogue as oppositional rather than collaborative, with little to differentiate dialogue from debate.

Dialogue as practiced today is quite different, engaging on many more levels than the narrow back and forth of argument. It is a zone of deep listening and powerful experiences, where collaboration toward group accomplishment leads to achievement in ways that competitive, adversarial discussions do not.

The National Coalition for Dialogue and Deliberation (NCDD), headquartered in Boiling Springs, Pennsylvania, is the US clearinghouse for all things dialogue. It's a network of more than twenty-seven thousand practitioners that provides resources and advice to individuals and organizations interested in using dialogue as a tool for public conversations. Its stated mission is to spread the practice of dialogue to discuss, decide, and take action on critical contemporary issues. Members include community leaders, public administrators, researchers, activists, teachers, and students involved in public engagement and conflict-resolution work. Encouraged by the NCDD, dialogue has become particularly popular with municipal, state, and federal levels of government striving to better connect with a public that is increasingly suspicious and mistrustful of politicians.

I talked with Sandy Heierbacher, who cofounded and directs the NCDD, about why dialogue has exploded as a means of public conversation. She cut her dialogue teeth working on race relations while a student in Vermont, where she realized that "you can't really have people sit and watch a presentation and have ah ha moments unless they're in a safe place where they can talk to each other about their backgrounds and how they came to have the sort of opinions they have."

I asked her why dialogue as a tool for public and private conversations has exploded in recent years. She paused for some time to consider, a signature element when dialogue is in the room, and then she responded: "As a culture, we have changed in certain ways that make it abnormal to have quality conversations with people we don't know or even people we do. We're so distracted these days with technology, going to the next thing. It's rare that you're fully present."

David Bohm, a theoretical physicist who was a key founder of the modern view of dialogue, saw dialogue as a solution to the twentieth-century malaise of isolation and fragmentation. He wrote: "Dialogue can be among any number of people, not just two. It is a stream of meaning flowing among and through us and between us, in the whole group, out of which may emerge some new form of understanding or shared meaning."

Participants read the room, perceiving currents in the situations around them, much as bees process a cascade of information into a dynamic understanding of the colony's condition and needs. Listening and total presence are essential. William Isaacs, a leading proponent of using dialogue in business, wrote: "The heart of dialogue is a simple but profound capacity to listen, requiring we not only hear the words, but also embrace, accept, and gradually let go of our own inner clamoring."

Dialogue involves focusing full attention on what others are saying by reducing our own inner racket. But good dialogue does not come only from paying attention to others, as important as that is. Responding constructively is a prerequisite for dialogue's back-and-forth.

It's very beelike; the multiple levels by which we engage through dialogue are suggestive of the bees' chemical and vibratory channels. Our most obvious senses are verbal and visual, and the subtleties of these communication modes are as nuanced and layered as those of bees in their hive. Presence and participation in human dialogues are also apparent in alert body language, increased ability to build on previous comments, and staying on topic.

The room slows down but feels more energetic. Silence ceases to be uncomfortable and becomes reflective and productive. Dialoguers hear each other in deep ways that build empathy and understanding, key components in a well-functioning dialogue environment.

Dialogue can happen in almost any context in which adversarial debate is the more common mode. Business meetings, classrooms, public consultation, and politically charged events all benefit from a dialogue approach. David Bohm provided a description of how a dialogue worked in a North American tribe in his book *On Dialogue:* "From time to time that tribe met in a circle. They just talked and talked and talked, apparently to no purpose. They made no decisions. There was no leader. And everybody could participate. There may have been wise men or wise women who were listened to a bit more—the older ones—but everybody could talk. The meeting went on, until it finally seemed to stop for no reason at all and the group dispersed. Yet after that, everybody seemed to know what to do, because they understood each other so well."

That's exactly what happens in a beehive. There's no leader, but a few individuals with more experience are listened to more than others. The queen is one of many, providing information but not otherwise dominating the conversation. Bees have no restrictions on who can contribute, and the conversation has no obvious starting or stopping point. But at some point it's apparent that the bees have understood each other because thousands of individuals coalesce into a remarkably functional unit.

At that point the bees are marvelously alert to each other, receiving and perceiving information important for bees, exuding a sense of awareness and intimacy as they act in tandem to accomplish tasks so that their colony will survive and thrive.

The conversation in the hive is much like story telling, with the bees accumulating information into coherent patterns that tell a tale. Perhaps the narrative is about the landscape through which they must fly to find nectar-bearing flowers or the story concerns the attacking wasps at the hive entrance, against which worker bees must mobilize, potentially sacrificing themselves to save their hive.

Whatever the message, listening is paramount. The connectivity provided by the bees' many communicative channels creates the colony's capacity for the collegial, collaborative environment that characterizes social insect colonies.

We too use story and listening to connect ourselves to the larger whole. In that way we share with honeybees that miracle of merging the individual with the collective through intense communication, which is so essential to a smoothly functioning community.

∘ ∘ ∘

How does bee time bring us into dialogue? We know when we've entered it: awareness heightens, the clock slows, all senses are captivated. The Canadian writer Robertson Davies described that state as "Where the stuff comes from, what happens to it, how the unconscious and the conscious must be allowed to kiss and commingle, and then how the conscious has to do the editing."

In dialogue we are at our most fully focused and attentive, present and in the moment. The outcomes and new understandings are sometimes experienced as lightning bolts of clarity, at other times as the slow lifting of a mental fog.

Insights happen. Perhaps we become aware of a new concept or more often put already-known ideas together in a new way, the world around us comprehended and experienced differently from the way we perceived it before.

Writers know the state of dialogue well—as a more internal process. The great British novelist Ian McEwan refers to it as "alert passivity." His character Clive in the Booker Prize–winning novel *Amsterdam* is a composer with the plum commission to write a symphony to usher in the new millennium, but he's blocked.

He decides to head north from London to clear his head by walking through the remote, rugged English countryside. He finds the notes he was seeking as the rhythm of walking takes over, and he hears a bird call: "His mind was contentedly elsewhere when he heard the music he was looking for, or at least he heard a clue to its form. A large grey bird flew up with a loud alarm call as he approached . . . It gave out a piping sound on three notes which he recognized as an inversion of a line he had already scored . . . He had no doubt that it was not a piece of music that was simply waiting to be discovered; what he had been doing, until interrupted, was creating it, forging it out of the call of a bird, taking advantage of the alert passivity of an engaged creating mind."

The state that artists call creativity or flow is akin to bee time. Artists find it too-often elusive and value it above all other tools. David Bayles and Ted Orland, in their slim but influential book *Art and Fear,* use the analogy of a novice going mushroom hunting with an expert to describe the muddle before an idea has crystallized and the clarity when you get it: "Go mushroom hunting with someone who really knows mushrooms, and you'll first endure some downright humiliating outings in which the expert finds all the mushrooms and you find none. But then at some point the world shifts, the woods magically fill—mushrooms everywhere!—and a view that was previously opaque has become transparent."

Business treasures the innovation and clarity found during dialogue, driven by the power of ideas to trigger new ventures and profit. The corporate lexicon describes dialogue as "brainstorming," "clean-sheet thinking," "lateral thinking," and "thinking outside the box." Creativity, novelty, and simply smart thinking are a premium product in the corporate world, whatever the buzzword used to describe the process. They provide the underpinning for a vast and lucrative network of consultants and strategic planners who assist business in achieving that creative state.

Neurobiologists have attempted to capture the states that are similar to bee time in brain images, hoping to nab a millisecond of creativity and inventiveness in a computer-generated picture of synapses firing. One study from Johns Hopkins

University put six jazz musicians into a magnetic resonance imaging (MRI) device, a scanner that illuminates areas of the brain responding to various stimuli. The MRI was modified to contain a keyboard, on which each musician first played a memorized composition note for note and then improvised while the machine imaged the performer's brain.

Results from the six musicians were remarkably similar. A broad portion of the front of the brain that extends to the sides (dorsolateral prefrontal cortex) slowed down during improvisation. This area has been linked to self-censoring, such as carefully deciding what words you might say at a job interview. Activity increased in the center of the brain's frontal lobe (medial prefrontal cortex), a region linked with self-expression and autobiographical storytelling.

The study's authors interpreted the images as snapshots of creativity, neurological underpinnings of the trancelike state jazz artists enter during spontaneous improvisation. The musicians were each telling their own musical story, shutting down impulses that might impede the flow of novel ideas and opening the channels where creativity dominates: bee time captured in a computer image.

If we could do a communal MRI of a honeybee colony's collective brain, I imagine it would reveal an open receptivity to information flow and a breakdown of any barriers to lateral thinking. The channels that process and receive odors and vibrations would present as neurological linkages that merge diverse stimuli into a coherent message, followed by brain commands for bees to act in ways that meet the colony's requirements.

This imaginary MRI would also reveal flexibility, with constantly changing information leading to quick improvisation and appropriate shifts in activity and workload. Our scanned bees would be quickly alert to changing stimuli, connected to each other and open to redirecting their workload as needed.

When we participate in the human equivalent of a colony conversation, dialogue, we experience something uncannily similar to the bees: the miracle of meshing with many others through the passive alertness of an engaged mind. Maurice Maeterlinck called this state of hyperawareness and collective behavior "the spirit of the hive" in his 1901 book *The Life of the Bee.* Like an iceberg, only the tiniest portion of the factors that determine our alertness is visible above the waves.

Studying bees is a powerful tool to tease out what lies beneath the surface because we can experiment and dissect phenomena in ways we can't for humans. In that way the bee world opens a window into understanding the seemingly incomprehensible complexity of how we communicate with and respond to each other as individuals and as a social collective.

Maeterlinck described the beauty of bee society in a way that is equally resonant for our own: "the enigma of intellect, of destiny, will, aim, means, causes; the incomprehensible organization of the most insignificant act of life."

11

Lessons from the Hive

🐝 Bees can be the richest of guides to the most personal understandings about who we are and the consequences of the choices we make in inhabiting the environment around us. Conversations with beekeepers about how they are affected by their time in the bee yard show a remarkable consistency. Words like "calming," "peaceful," and "meditative" come up over and over again, and beekeepers visibly relax when talking about their bees.

Beekeepers exhibit an emotive connection with their bees, a passion, a deep and abiding friendship, a layering of human emotion onto a species that, unlike household pets, is unlikely to respond in kind. It's a one-way experience, this relationship with bees, but no less powerful for being unilateral.

Dave Hackenberg, the commercial beekeeper from Pennsylvania who was the first to report colony collapse disorder, may be the world's most stressed beekeeper as a result of the ongoing challenge of maintaining his business and the non-stop requests for interviews from the media. Still, after keeping bees for fifty years, Hackenberg finds his bees therapeutic: "There's a lot of things to learn from bees. If you're a beekeeper looking at beehives, you stop and pay attention to what's going on—you can learn some valuable lessons. It's a good way to get rid of your stress."

Bob Wellemeyer, the honey judge from Virginia, considers bees to be almost addictive in their ability to stimulate relaxation. When he has a bad day, he says, "Once I get into the bee yard, the world goes away. It's meditative. I just get in with the bees and I'm at peace."

I asked Brian Marcy, a hobby beekeeper and vice president of the Montgomery County, Pennsylvania, Beekeeping Association, what lessons he's learned from bees. I could see him mentally entering the apiary before he responded: "Patience. If you're not patient with your bees, you'll get stung, they'll let you know it. Also to be very aware of what's happening around you, to focus and go slowly, not to rush. It's been a calming experience for me, gives me a chance to step back and really focus on nature."

That calming patience with bees has spilled over into more than his demanding work life. His twin brother is facing stage-3 cancer, and Marcy's found that his "ability to focus on and be with the bees has really helped me to go through those periods when I know that my brother is struggling."

The way beekeepers describe their experiences seems more appropriate for a yoga class or the study of Zen Buddhism than being around an insect that could sting the living daylights out of you. Yet, a tacit understanding exists between beekeeper and colony; if you're calm around honeybees, they will be calm as well, creating a dynamic that feels to the beekeeper like a relationship.

A part of that interaction results from beekeepers perceiving and internalizing the focus and awareness that characterizes how bees relate to each other. Honeybees are incessantly aware of what is going on around them, persistently receiving information and returning comments through chemical, visual, sound, and touch modalities.

That sense of presence radiating throughout a honeybee colony creates a similar sense of heightened awareness and participation for beekeepers in the apiary. Intense absorption translates into the calm feeling so often recounted by beekeepers when describing how they experience their bees and persists when we leave the bee yard and return to our more human habitats.

Children are particularly receptive, quickly turning their fear into curiosity. I talked with Nicky Grunfeld, a university student who did an internship with the Environmental Youth Alliance developing a curriculum that uses bees to engage young children with agriculture. She described one particularly riveting moment: "Something else I've learned from bees, especially working with children, is how little it takes for something that is seen as provoking so much fear to disappear quickly and turn to awe. Last summer a swarm landed in a tree a little distance from the hive; I got to go with the children and gently place our hands on the swarm, because the bees are quite docile at that point, and one of the kids commented that 'it feels like you can feel their heartbeat.' It was a really special moment."

Beekeepers don't always start out feeling calm. For beginners, beekeeping often means overcoming your own fears. Suzanne Matlock, a Philadelphia hobby beekeeper, described the moment when she finally lost her terror of being stung: "In the beginning I was afraid to put my hands into the beehive and I weaned myself off of wearing gloves over some time. It took about three years. I got thinner and thinner gloves until I had no gloves. But the moment I lost all fear [was when] a little kid wanted to learn about beekeeping and come close to the hive. I didn't want him to miss that chance, so I took my suit off and gave it to him, and I just picked up the frames without gloves. It was wonderful. That was my moment. I learned that they're very calming, which is something I wouldn't have guessed."

Another common lesson beekeepers report is an enhanced capacity to notice things around them, especially the subtle markers that may be below our conscious perception but so often define situations. It's the value of small things, perceived when we slow down and pay closer attention.

Nicky Grunfeld described how her experience with hives radiated out from the apiary: "It's not only at the hive. I walk around, check the flowers, and see if there are bees collecting pollen. I begin to notice more. It enriches everything. It's taught me to slow down and appreciate the little things."

o   o   o

Our fate is entangled with that of the bees. There have always been lessons to be learned from bees both wild and managed, but our need to pay closer attention to the apian world has become more compelling in recent years.

We can't afford to ignore what the study of bees can tell us about our increasingly tenuous affiliation with nature. We are a species that sees our prosperity, and often our very survival, as depending on our ability to control nature by managing its fields, forests, and waterways for our own purposes, creating separation between us and other organisms like that between master and servant.

To enjoy and appreciate or to exploit and harvest: those are the extremes of our dilemma. We too often err on the side of economy at the expense of ecology. Bees tell us we have not yet found that sweet spot where we can benefit from our ecosystem without decimating nature. Wild bees are especially captivating in deepening our appreciation of the economic services that ecosystems can provide given an opportunity. They present compelling evidence that sustainable agriculture is both practical and achievable.

Claire Brittain, the UC-Davis researcher who kindly showed me around the almond orchards of central California, has become acutely aware of the ways bees interact with their environment. When she works in the almond groves, she says, "I look at the bees and think, those bees are important right now, for that crop, but the flowers before and right after are important to those bees and the wider landscape. Even though you're looking at something really small, everything is connected."

Bob Wellemeyer learned a similar lesson as a nine-year-old beginning beekeeper: "I became more aware of floral sources blooming and how their lives interact with ours. I started looking at books and lists of floral sources a mile long and blooming times. Before that I went through life not knowing that trees had flowers or even produced honey. There's a whole other world out there when you start looking around."

Ecosystems may be complex, but their intricacy is composed of very simple and straightforward interactions. For bees and plants, their highly coevolved relationship distills down to one simple point: they depend on each other. Bees need abundant and diverse flowers to provide them with nectar and pollen, while flowers require bees to move between flowers and pollinate.

Nowhere is that codependence better articulated than in agriculture, where the additional layer of human economy is superimposed on the natural order of bees and flowers. Farming at its best recognizes that relationship and weaves diverse and healthy habitats together with the bees. But at its worst, agriculture is a death sentence for wild and managed bees alike, to the detriment of both nature and us.

Studying bees in the large-scale farming habitats of central California has particularly affected Claire Kremen from the University of California at Berkeley:

"I think the most profound lesson I've learned from bees is about how really awful our food system is. The dominant model for food production is this industrialized model that is really bad for wild bees. The system produces bloom for only a couple of weeks a year; during the rest of the year it's pretty barren, no crops blooming, no weeds blooming. I started realizing what a crazy way of producing food this is. In these hot, dusty, very barren landscapes, I was struck with how unpleasant they were to be in. We found very few wild bees in these intensively cultivated agricultural fields. It was very, very clear what a negative relationship they were having."

The lesson from spending time with bees in any agricultural setting is clear: mess with pollination and both bees and crops will suffer. And suffer they have, with wild bees deprived of nesting sites and both wild bees and managed honeybees confronted with few forage choices due to vast single-cropped acreages and highly effective weedkillers. Further, bees in and around farms are exposed to myriad insecticides and fungicides that kill them directly or reduce their immune-system functions so that they are more susceptible to diseases and agricultural chemicals. For crops, yields are reduced by a lack of pollinators, farmers' profits decline, and consumer prices increase as honeybee colonies become scarcer and more expensive to rent.

The solution is easy to envision and feasible to implement. We need to value bees as ecosystem services and provide healthy habitats that would be of benefit to both wild and managed bees.

The mechanisms for change are clear and, primarily involve reducing pesticide use and diversifying farms. These steps would enhance our own food security and provide many benefits beyond helping bees. The challenges are not so much practical as a matter of changing mindsets since the action items proposed by Kremen and others are economically realistic and the details of how to proceed are well researched. Success will come only if we move away from our current manage-and-control attitudes and toward strategies where we are more aligned with nature as an ally. The idea that ecosystems could provide the services of bees and other beneficial organisms runs counter to current practices, where pest control is chemical and pollination is primarily through rented honeybees. Perhaps the current crisis with pollination will stimulate broader thinking about how we grow food and what it means to be sustainable.

Bees provide other examples of how we can readjust out relationship with the land and contribute to healing other long-standing environmental conflicts. Perhaps no scar on the earth is more visibly noticeable or has generated more protest than the devastated land left from strip mining in Kentucky, but even here bees are providing a way forward to heal the land and the relationships between concerned citizens and the coal companies.

Like agriculture, coal mining is a hot-button issue in which economy and environment collide, but Tammy Horn and her project, Coal Country Beeworks, are providing a way to improve the land while diversifying livelihoods outside of coal mining.

Horn herself is Kentucky-born and after university returned as an English professor. She learned beekeeping from both sets of Kentucky grandparents, which inspired her research into a book about bees in the United States and then one on women in beekeeping.

Her interests gradually shifted toward Appalachian studies, and her current appointment at Eastern Kentucky University's Center for Development, Entrepreneurship, and Technology has allowed her to implement Coal Country Beeworks. She's collaborating with four coal companies to replant surface mine sites with nectar-producing trees, including tulip poplars, black locusts, and sourwoods, as well as ground-covering wildflowers.

The project is based on the same principle as Kremen's vision for California agriculture: economic diversity depends on landscape diversity. And it's not without its challenges. It took some time for the beekeeping and the corporate sides of this project to trust each other. Horn told me: "When I first started working with coal companies, it looked like the project wouldn't go forward, and I had to suppress a position that would have been a very selfish one. I am from Harlan County, and it was really important that this move forward. I had to learn the importance of not holding tightly to a position if I wanted to make a contribution to my region, to my people."

Whether it is coal mining in Appalachia or farming in California, bees can serve as a bridge reconnecting us with land that has been modified beyond recognition. Yet, in these and many other habitats, enough remnants of original ecosystems remain to enable us to begin reconstructing habitats that provide a better balance between ecosystem integrity and human economy.

Beekeeping also connects us with the land through our association with the honey each region produces and the beekeepers whom we come to know through their product. Our enjoyment of food is enhanced by the stories and personalities from which it comes. Appreciating a carefully produced artisan product like honey is a counterbalance to the distance mass agriculture creates between us and the sources of our food. In that way bees expand our experience of each other and the world around us.

A final environmental lesson we can learn from honeybees comes from exploring the boundary between feral and managed. Honeybees inhabit an unusual dimension in agriculture, with the managed bees in colonies being closer to the feral version than any other domesticated species. Honeybees remain essentially wild, and even the hives we build for them are quite similar to the nests they construct in nature. There has been very little selection on bees, and swarms that leave managed hives and find a nesting site in a hollow tree do quite well without our intervention. In contrast, cows, pigs, chickens, and other domesticated animals usually fail quickly on their own.

This situation of managing an essentially feral species provides an unusual window into how much we have changed the world around us. Ironically, honeybees, feral as they are, have fallen victim to excessive management by beekeepers and to the highly manicured and managed agricultural environments in which they are often kept.

Bees in cities provide a counterbalancing example of how to better integrate the feral and the managed. Cities have become healthier habitats under contemporary strategies of urban planning; wild bees have become more abundant and beekeeping with honeybees more feasible as cities have increased green spaces and promoted urban agriculture.

Perhaps that lesson from cities would work as well with farmland. Perhaps we can learn something from an unlikely teacher, urban bees, about reinvigorating our relationship with the land.

o o o

We are drawn to honeybees for many reasons, but their social behavior is particularly riveting. We, and honeybees, represent two pinnacles of sociality among the earth's creatures, and we can learn much about ourselves by observing them.

What is particularly compelling about bee societies is how well they function, especially how seamlessly individual and collective interests merge. We have much in common with honeybees, but if there is a single striking difference, it's in how much more we humans struggle to find the appropriate balance between personal gratification and societal imperatives.

We are fascinated by honeybee sociality in part because honeybee workers seem to achieve aspects of collective function that we aspire to but do not always achieve. Much of their success emerges through hard work and a capacity to submerge individual aspirations into the collective of the hive. We strive for the altruism, work ethic, flexibility, teamwork, and communication, which are highlights of honeybee colony life, often failing to reach what honeybees seem to do easily.

Attention to honeybees can inform our own choices about how we can each best integrate into the human colony. Almond growers and beekeepers Drew and Tyler Scofield are great admirers of how honeybees sacrifice for the common good: "Their organization blows my mind, the social order of things," says Tyler. "They put the colony before themselves, they're the most selfless organism I've ever seen, completely dedicated to the colony," adds Drew.

This sense of altruism is not absent in human life. We celebrate the soldier who throws himself on a grenade to save his comrades, the nun who establishes a soup kitchen for the indigent, the unmarried daughter who sacrifices her future to care for her aging parents.

But these altruistic acts are balanced by a streak of selfishness that is also part of our humanity. We may celebrate acts of selflessness, but we also reward the kind of selfishness that makes hedge-fund managers billionaires. We are attracted to both poles in our human character, the narrow self-interest promoted by an Ayn Rand and the personally punishing charity of a Mother Teresa.

Honeybees have no such ambiguity; with them it's always colony first. There's no income inequality in the bee world, no crime in the colony, no awards or status symbols to assess one bee's success as greater than another's. Conflict exists, but it's managed by a social system that favors collaboration over individual gain.

We are not honeybees, and spending time in apiaries is not going to eliminate our own tension between the personal and the communal. But we can be inspired by the honeybee's example of sacrifice, inducing our own reflections on where we wish to place ourselves on the individual-to-collective continuum.

Honeybees also focus our attention on work: how we perform it and the value we place on it. Worker bees are pervasive as symbols of righteous toil in religious, civic, and literary representations dating back many thousands of years in almost every human society. We consider honeybees to be models of efficiency in the ways they collaboratively delegate the many tasks colonies must perform to survive and thrive. But we also ascribe virtue to their labor, an aspiration unlikely to be found among workers in a beehive but one that permeates much of our human thinking about work.

The irony of our characterization of honeybees as hard workers is that they spend a majority of their time not working. Rather, worker bees create a balance of work and rest to ensure individuals have the strength and colonies have the balance to allocate work when and where needed. In a modern world, where work-life balance is seen as a challenge, we might be well served to learn that lesson of moderation rather than promoting the "busy as a bee" idea.

Beekeepers often mention learning about flexibility from their hives, a message that resonates from the adaptable way bees move between tasks during their lifetimes. We've seen how worker bees have a natural progression from within-hive to outside jobs during their thirty or so days of summer life, and how that progression can shift depending on colony needs at any given time.

That idea of moving through our own lives with suppleness is yet another lesson we can learn from the hive. Kentucky beekeeper Tammy Horn credits her honeybees with influencing her professional shift from English to designing bee-friendly landscapes: "I think that in this career transition, what I learned from beekeeping was the importance of being flexible, of being able to change duties and obligations as I grew up. People think that success is a linear process, but it's not. You have to work at it every day."

Colonies also exhibit a sense of order and logic, in which many individual decisions accumulate into highly complex societal functions. Jim Bobb is the current chair of the Eastern Apicultural Society and a beekeeper whose life has embodied the theme of flexibility. He grew up on a farm outside Philadelphia that served as George Washington's headquarters in 1777 and went on to found a software company by day while owning and managing a Brazilian nightclub by night in San Francisco.

He has returned to his Pennsylvania roots and now keeps bees in a number of local public gardens. His engineering background has helped him appreciate the order that permeates honeybee colonies and to apply that sense of orderliness to our world: "In watching the bees it's just fascinating how everything fits together. The bees have such logic to them; everything just makes sense in the way the bees all work together. You try to apply that to what you're doing in life—you're trying to get teams to work together, and somehow in the beehive all of those creatures work together. It's something that we should be able to do ourselves."

We strive for that sense of societal integrity where the sum of our personal endeavors creates a smoothly functioning whole much more coherent than its individual parts. A bee colony is the ultimate expression of teamwork and collaboration; observing bees can be a useful proxy to learn how we, too, can cooperate more effectively.

If there is a single element that stands out to explain why bees work together so well and from which we can learn some valuable lessons, it's their intense communication with each other. Honeybees excel at exchanging information with and maintaining a continual awareness of the hive mates around them, an area where we could do considerably better.

Bees listen to each other, deeply, all channels on, using every mode of communicating we know of and probably some we're not yet aware of. Vision, odor, taste, hearing, touch, vibration, magnetism, electric fields—the input is constant and the interactions intense.

Honeybees also appear endlessly curious, hungry for information, soliciting news from each nest mate they encounter and, in turn, passing knowledge back. These exchanges can be seen in a flurry of activity, bees stroking each other with antennae and legs, mouthparts exchanging food and pheromones.

In the field, worker bees scour the countryside seeking nectar and pollen, communally observing the territory for miles around the nest. When a find is discovered, individual scouts are quick to return and dance, recruiting foragers by spreading the knowledge accumulated on their scouting expeditions.

The curiosity and listening skills of honeybee workers stand as beacons to those of us who aspire to being more fully present to the world around us. Honeybees live very much in the moment, and their example reminds us that attentiveness is a key tool for successful teamwork.

Honeybees have achieved what many of us strive for: a life lived in the moment, replete with deep, substantive interactions, enriched by relationships with others and a profound connection with the environment around us.

o   o   o

Honeybees teach us something else. Best described as inspiration, it's expressed in the way that the magical, mystical essence of the hive inspires our creativity and wonder. The most powerful translators of the marvels presented by bees may be artists, who provide distinctive insights through various media.

The ancient Greeks considered the muses to be nine goddesses who were the inspirations for creativity in science, literature, and art. "To muse" means to be absorbed in thought, to ponder and contemplate; immersion in bee time seems to invoke the gods and goddesses as we mortals meditate upon and attempt to express the wonder inspired by bees.

Aganetha Dyck, whose art collaborates with honeybees, has given considerable thought to the inspiration and interaction she gets from bees. I asked her about bees as a muse for imagination, and she responded in an e-mail: "Honeybees have and continue to inspire designers, architects, and artists throughout the world. Honeybees inspire me to work with them because they feed my curiosity; they allow me to experiment, teaching me to think into their box and out of my box. The bees are my muse because they allow me to contemplate, to wonder and reflect on who is this truly mysterious, magical, gift giving pollinator."

For dancer and choreographer Gail Lotenberg, inspiration from bees has come from the interplay of activity and quiet: "Bees have such a well-known frequency or rhythm connected to them and their movements and their sonic trademark. I was entranced by the invitation to ramp up to their frequency. But then there is the beautiful contradiction, that to

be safe around bees, you have to actually move toward still-
ness. This contradiction is rich territory for imagination and
inspiration."

For quilt maker Hope Johnson, her craft is inspired by the
geometry of the hive. She runs Decorative Artworks in Shel-
burne, Vermont, specializing in quilts that show the inside
of hives through exquisitely stitched details of comb, larvae,
adult workers, nectar, and pollen.

Her quilts are composed of seemingly infinite stitches and
detail, each quilt taking more than four hundred hours to
complete. They are very much like impressionist paintings,
where a close look reveals thousands of tiny dabs of paint
that resolve into images when the viewer steps back.

She described the interaction between geometry and inspiration in her work: "The honeycomb architecture is an elegant example of natural geometric hexagonal tiling. The challenge for me is to render the macro (whole picture) and the
micro (details) in a design that has overall beauty and compo-
sitional integrity. The example set by the bees of effort and
persistence is an awesome accomplishment of industry, cooperation, and instinct that keeps me encouraged and focused
on my work."

Besides the bees' example of hard work, the other aspect of
honeybees that inspires Johnson during the tedium of quilting
is that "bee society, behavior, and biology have challenged me
to represent their beauty, societal hierarchy, and order. If artists are more tuned in to sensory experience, then bees have it
all: the music of buzzing, the drama of pollination against a
backdrop of floral chroma, the dances of flight and communication, the architecture of the hive, the aroma of beeswax,
and the culinary delight of honey. Bees are all-round creatures
of light and as such are a very appropriate muse for artists."

Artists have found outlets to express what many of us
feel about honeybees. They possess an essential spirit, one

much more expansive than the sum of the isolated behaviors that so impress—behaviors that come together to make up the colony.

Hope Johnson spoke of the essence of bees this way: "If there is a spiritual aspect to bees, then it is in their 'bee-ness.' As a result of creating with bee imagery, I have developed connection and community as well as artistic purpose and gratitude. They are a constant source of wonder and kindling for the imagination."

Bees increase the range of our imaginings, augment our attention, and inspire a sense of awe. It's the world that Aganetha Dyck appreciates: "Spending time in the honeybee's world opens a new world of wonder for me. Being with the honeybees makes my ordinary life stand still and makes time disappear. Opening a hive for the first time is similar to traveling to a strange, new country. The sights and scents, the sounds and warmth, the movement—these are rare discoveries."

o o o

Bees, honey and wild, were very much with me while writing this book, but I was also inspired by the beekeepers and bee aficionados I encountered at every turn. The human world that adjoins bees is a rich one, full of characters and stories representing quirky slices of human nature.

Mostly I was taken with the dedication and passion of the personalities who came to populate my bee time. I was reminded that working with bees serves as a substrate upon which to build human achievement and satisfaction, a canvas upon which to paint our interests and ambitions.

Alice and Désirée come to mind, the high school students whose passion for bees ignited the "Once upon a Bee" project. Their ambition was to learn everything they could about wild bees in the city and use that knowledge to express their intense personal dedication to environmental stewardship. Their craving to have a positive impact on the world stimulated municipalities, science museums, garden groups, and many individual citizens to implement bee-preserving habitats throughout the region. Bees and the flowers they pollinate are much healthier for their efforts.

Or Hives for Humanity, the beekeeping program in Vancouver's poorest district, which provides job and life-skills training for community members whose lives have been challenged by poverty, addiction, and mental illness. To see the sense of calm that descends when individuals who are generally agitated are around bees, and their pride in producing high-quality honey, demonstrates how bees can have considerable positive impact on the lives of their keepers.

Possibly the most unusual group I encountered were the shamanic beekeepers who imagine a mystical presence in and around their hives. But behind their New Age language is their connection with the interactions that make colonies so intriguing and the pollinating functions that are the thread lacing together healthy ecosystems. Shamanic metaphors may be difficult to relate to for those of us with a practical bent, but we share the abiding sense of wonder about bees, for which language is often inadequate.

Not everyone I met was commendable. Honey adulterators and importers shipping and selling tens of millions of dollars' worth of tainted honey around the globe come to mind. Still, their crimes were balanced by the many artisan beekeepers I came to know, individuals who care deeply about the quality of their product and its connection to the land they love.

Much of my time was spent with farmers, evaluating the current crisis facing honey and wild bees alike. I never encountered a producer with ill will toward bees, but growers are trapped in a system that pushes farming toward an industrial scale with heavy input of chemicals and fertilizers. We can do better, although it's going to take systemic change to overcome the inertia of contemporary agriculture. But change we must, or the next generation of farmers may not have the bees to provide essential environmental services.

No matter how extensively we have modified our world for human purposes, we still crave a bond with nature and with each other. We may harm our environment through careless development, but we also strive to save and protect the other species with which we share an increasingly fragile planet.

Bees act as connectors to acquaint us with our neighbors and stimulate deep collaborations and friendships. Their sociality and the complex environmental web bees inhabit provide a muse that guides us in reflecting on who we are and want to be, with each other and with the world.

That's why bee time is so compelling: As we come to know bees, we see an echo of ourselves.

Epilogue

Walking out of the Apiary

I began walking out of the apiary in 2002, for reasons most human: it was time for a change.

My exit from bees was slow. I wound my research laboratory down and graduated the last of my students over the next four years, closing the lab doors officially in 2006. I had moved laterally within Simon Fraser University (SFU) to establish and direct the new Centre for Dialogue. The demands of that position simply left no time to adequately supervise my students, apply for funding, conduct research, or attend scientific conferences and beekeeping meetings.

I entered into this transition willingly, enthusiastic for the new challenges at the Centre, but it wasn't as easy to walk out of the apiary as I had naively imagined.

I moved on from bees for many reasons, but perhaps the most compelling was an affliction of middle age: restlessness. The challenges of research were becoming routine, the constant applying for grants was wearing, and teaching the same courses monotonous. My department was asking me to lecture to larger and larger classes, when my interest was in sitting down with just a few students and listening, mentoring rather than instructing.

The university had been given a stately old 1920 neoclassic bank building in the heart of downtown Vancouver and renovated it exquisitely for a new purpose, a center for dialogue. The heart of the building is its 154-seat round hall, with tiered circular rows designed for interaction and communication. The intention was to hold dialogues rather than debates or formal talks. But in the excitement of renovating, SFU hadn't made provisions for programming to fill the Centre and really hadn't grappled with what it means to dialogue, for students, for faculty, or for the community.

We first began developing the concept and practice of dialogue with students, creating an unusual one-semester, intensive program designed to inspire students with a sense of civic responsibility and encourage their passion for improving society. Each semester the program offers an original experience that links classroom with community and creates space and time for students to reflect on what they are doing and why it matters.

The Semester in Dialogue is based on three principles woven together into a learning tapestry. First is dialogue, respectful listening that builds deep relationships through free expression of views and open-minded exploration of differences. The second is experiential education, emphasizing learning through doing. Students apply skills to community issues, reflect on the results, and use this learning to provide context for who they want to become in the world. Engagement is the third pillar on which our program rests, developing partnerships to enhance the well-being of communities. Students engage the community in a variety of ways, including partnering with organizations on projects, convening dialogues to address pressing issues, and hosting leaders from all sectors and diverse points of view in the classroom.

We soon expanded beyond teaching into community programming, and the Centre for Dialogue has become an important resource to generate nonpartisan and constructive communication on difficult topics. We work with government, business, and nonprofit groups to explore critical issues that affect the social, economic, environmental, and cultural well-being of our city and province, as well as nationally and internationally.

A profound belief in the power of dialogue guides our work, including the ability of dialogue to humanize participants and bridge understanding through authentic inquiry and exchange. Mutual curiosity, collaborative inquiry, and compassionate listening act as alternatives to adversarial exchange and encourage transformative social change.

The breadth of issues and types of programs we generate are perhaps best understood by a quick scan of recent events. Our students put on a public dialogue, "Building Community, Building Health," bringing together health practitioners and citizens to reimagine health from a community perspective. The Centre held a series of public dialogues on the issue of mobility pricing, a hotly discussed proposal in our region to charge automobile drivers based on where, when, and how far they drive. In the winter of 2014 we held five programs focused on reconciliation among aboriginal inhabitants and the pluralistic settler population of Canada, including a poetry reading, an event for high school students and their teachers to explore reconciliation in the schools, and a public dialogue exploring how peoples of many cultures have resolved injustices.

The city of Vancouver convened a workshop with us on flood-proofing, developing practical policies to deal with the real likelihood that storms and rising sea levels caused by climate change will affect urban areas. We held a month of one hundred kitchen conversations across British Columbia about the future of our provincial economy, culminating in an economic summit notable for broad agreement among its diverse participants, ranging from right to left wing, prosperous to poverty stricken, on how to combine economic prosperity with environmental and social sustainability.

While these and the many other issues we probe at the Centre have little, if anything, to do directly with bees, navigating the politics of beekeeping and agriculture proved to be an excellent training ground for engaging with controversial public issues. The lessons learned from bees provide a rich set of principles and practical methods to bring to the dialogue table.

For one thing, issues associated with bees and agriculture can be highly contentious. Working with bees and beekeepers created innumerable opportunities to consider broader societal issues such as pesticide use, genetically modified crops, and farm policy, all issues impressive in the wide range of opinions and positions held by stakeholders and the public. I learned that listening and balance were considerably more effective than disputing, and solutions more easily reached by understanding diverse points of view than by walking into rooms ready to argue.

My time with bees also ideally predisposed me toward the mindset of dialogue. As I reflected on what to apply from bees into my new dialogue world, I realized how much my life outside the apiary has been enriched through bee time.

I learned perseverance from watching bees work and assimilated from the bees that engagement requires persistence for success. Taking care of honeybees made me more responsible; watching how they work for their colonies expanded my own capacity to care for others. Family and community became integral to my life, for which bees deserve major credit.

Focus, engagement, and presence matured as I observed those qualities in bees. I began to listen more carefully, paying attention to what people said and the meanings beneath their words, maturing my capacity to dialogue. The intrinsic collegiality of bees grew my own collaborative nature, with the realization that as individuals we can be most effective when cooperating with others.

Another thing I drew from bees that crossed over into my dialogue world was appreciation for how diverse perspectives contribute to the most stimulating and productive interactions. But appreciation is lost if not expressed, so we developed a simple thank-you for our participants that grew from my research laboratory's practice of giving honey to anyone who contributed to our endeavors.

When we closed our apiaries in 2006, I stockpiled hundreds of jars of our last harvest of Heavenly Honey and continued my earlier tradition of dispensing them as gifts to the speakers and leaders who came into our dialogues. Each jar is passed on with our profound thanks, and both the gratitude and the honey are much appreciated.

We gave away the last of the Heavenly Honey jars this year but will continue the tradition by giving each guest a jar of Hives for Humanity honey, a physical reminder that dialogue must include the most disadvantaged citizens if it is to be meaningful.

Leaving the apiary behind was not nearly as easy as I thought it would be, much as I have thrived at the Centre. I find myself missing the bees and the array of friends and colleagues bees brought into my world. I've also been increasingly disturbed by the dire state of bees and beekeeping today, which has degenerated considerably since I moved over to the Centre for Dialogue twelve years ago.

Missing the bees and deep concern about their current plight grew into this book, a whisper at first but eventually more insistent, a way to reconnect with bees and beekeepers, my humble expression of gratitude to the bees for all they have given.

Whatever comes next, I'm sure it will continue to be richly flavored by the time I spent among the bees.

References

The references here, by chapter, provide key sources, citations for any printed or digital material quoted in the text, and addresses for some of the websites that I found particularly useful. Although not comprehensive, it provides a solid entry point for readers interested in exploring these topics further. If you'd like more general information about honeybees, see my book *The Biology of the Honey Bee* (Cambridge, MA: Harvard University Press, 1987). For updates on bees or dialogue please visit the following websites:

Personal: www.winstonhive.com
Centre for Dialogue: www.sfu.ca/dialogue
Semester in Dialogue: www.sfu.ca/dialogue/semester

1. BEGINNING WITH BEES

Cappellari, S., H. Schaefer, and C. C. Davis. 2013. Evolution: Pollen or Pollinators—Which Came First? *Current Biology* 23: 316–318.

Crane, E. 1999. *The World History of Beekeeping and Honey Hunting.* London: Routledge.

Huber, F. 1806. *New Observations on the Natural History of Honeybees.* http://www.bushfarms.com/huber.htm#letter1.

Ollerton, J., R. Winfree, and S. Tarrant. 2011. How Many Flowering Plants Are Pollinated by Animals? *Oikos* 120: 321–326.

Vanbergen, A. J. 2013.Threats to an ecosystem service: pressures on pollinators. *Frontiers in Ecology and the Environment*, 2013; 130422054656003 doi: 10.1890/120126.

2. HONEY

Ames Farm. http://www.amesfarm.com/.

Food and Agriculture Organization Global Statistics. http://faostat.fao.org/site/573/DesktopDefault.aspx?PageID=573#ancor.

Food and Drug Administration. 2010. Indictment. http://www.fda.gov/ICECI/CriminalInvestigations/ucm226410.htm.

Fredericksen, B. 2008. Ear to the Ground. Land Stewardship interview (July 29). http://landstewardshipproject.org/posts/podcast/224.

Gibran, K. 1923. *The Prophet.* Eastford, CT: Martino.

Kwakman, P. H., A. A. te Velde, L. de Boer, D. Speijer, C. M. J. E. Vandenbroucke-Grauls, and S. A. J. Zaat. 2010. How Honey Kills Bacteria. *Journal of the Federation of American Societies for Experimental Biology* 24: 2576–2582. www.fasebj.org.

Leeder, J. 2011. Honey Laundering: The Sour Side of Nature's Golden Sweetener. *Toronto Globe and Mail* (January 5). http://www.theglobeandmail.com/technology/science/honey-laundering-the-sour-side-of-natures-golden-sweetener/article562759/?page=all.

Mavric, E., S. Wittmann, G. Barth, and T. Henle. 2008. Identification and Quantification of Methylglyoxal as the Dominant Antibacterial Constituent of Manuka *(Leptospermum scoparium)* Honeys from New Zealand. *Molecular Nutrition and Food Research* 52: 483–489.

Molan, P. C. 2001. Manuka Honey as a Medicine. *Proceedings of the Global Bioactives Summit, Hamilton, NZ.* http://bio.waikato.ac.nz/pdfs/honeyresearch/bioactives.pdf.

——. 2006. The Evidence Supporting the Use of Honey as a Wound Dressing. *International Journal of Lower Extremity Wounds* 5: 40–54. http://researchcommons.waikato.ac.nz/handle/10289/229.

——. 2011. The Evidence and the Rationale for the Use of Honey as a Wound Dressing. *Wound Practice and Research* 19: 204–220.

Moore, J. C., J. Spink, and M. Lipp. 2012. Development and Application of a Database of Food Ingredient Fraud and Economically Motivated Adulteration from 1980–2010. *Journal of Food Science.* doi:10.1111/j.1750-3841.2012.02657.x.

Moskowitz, D. 2004. The Taste of Here—Ames Farm Revolutionizes the Gathering of Honey, to Flabbergasting Effect. http://www.citypages.com/2004-11-03/restaurants/the-taste-of-here/ (November 3).

Mountain Honey. http://www.mtnhoney.com/index.htm.

National Honey Board. http://www.honey.com/.

Schneider, A. 2011. Asian Honey, Banned in Europe, Is Flooding U.S. Grocery Shelves. *Food Safety News* (August 25). http://www.foodsafetynews.com/2011/08/honey-laundering/.

——. 2011. Questions about the Quality of Imported Honey. *Food Safety News* (December 27). http://www.foodsafetynews.com/2011/12/the-7th-most-important-food/.

——. 2011. Tests Show Most Store Honey Isn't Honey. (November 7). http://www.foodsafetynews.com/2011/11/tests-show-most-store-honey-isnt-honey/.

Soffer, A. 1976. Chihuahuas and Laetrile, Chelation Therapy, and Honey from Boulder, Colo. *Archives of Internal Medicine* 136: 865–866.

Somal, N. A., K. E. Coley, P. C. Molan, and B. M. Hancock. 1994. Susceptibility of *Heliobacter pylori* to the Antibacterial Activity of Manuka Honey. *Journal of the Royal Society of Medicine* 87: 9–12.

Winston, M. L. 2004. Two Bottles of Mead. *Bee Culture* (February), 15–16.

3. KILLER BEES

Pimentel, D., R. Zuniga, and D. Morrison. 2005. Update on the Environmental and Economic Costs Associated with Alien-Invasive Species in the United States. *Ecological Economics* 52: 273–288.

Solman, P., and T. Friedman. 1982. *Life and Death on the Corporate Battlefield: How Companies Win, Lose, Survive.* New York: Simon and Schuster.

White, W. 1991. The Bees from Rio Claro. *New Yorker* (September 16), 36–60.

Winston, M. L. 1992. *Killer Bees: The Africanized Honey Bee in the Americas.* Cambridge, MA: Harvard University Press.

Zuckerman, E. 1977. The Killer Bees. *Rolling Stone Magazine* 244 (July 28): 50–57.

4. A THOUSAND LITTLE CUTS

Alaux, C., J. Brunet, C. Dussaubat, F. Mondet, S. Tchamitchan, M. Cousin, J. Brillard, A. Baldy, L. Belzunces, and Y. LeConte. 2010. Interactions between Nosema Microspores and a Neonicotinoid Weaken Honeybees *(Apis mellifera). Environmental Microbiology* 12: 774–782.

Aufauvre, J., D. G. Biron, C. Vidau, R. Fontbonne, M. Roudel, M. Diogon, B. Viguès, L. P. Belzunces, F. Delbac, and N. Blot. 2012. Parasite-Insecticide Interactions: A Case Study of *Nosema ceranae* and Fipronil Synergy on Honeybee. *Scientific Reports* 2: 326. doi:10.1038/srep00326.

Cornman, R. S., D. R. Tarpy, Y. Chen, L. Jeffreys, D. Lopez, J. Pettis, D. vanEngelsdorp, and J. Evans. 2012. Pathogen Webs in Collapsing Honey Bee Colonies. *PLoS ONE* 7(8): e43562. doi:10.1371/journal.pone.0043562.

Environmental Working Group. http://www.ewg.org/foodnews /summary.php.

Goulson, D. 2013. An Overview of the Environmental Risks Posed by Neonicitinoid Pesticides. *Journal of Applied Ecology* 50: 977–987.

Hawthorne, D. J., and G. P. Dively. 2011. Killing Them with Kindness? In-Hive Medications May Inhibit Xenobiotic Efflux Transporters and Endanger Honey Bees. *PLoS ONE* 6(11): e26796. doi:10.1371/journal.pone.0026796.

Johnson, R. M., H. S. Pollock, and M. R. Berenbaum. 2009. Synergistic Interactions between In-Hive Miticides in *Apis mellifera*. *Journal of Economic Entomology* 102:474–479.

LeConte, Y. 2010. Interactions between Nosema Microspores and a Neonicotinoid on Honey Bees. Presentation at the American Bee Federation meeting, Orlando, FL. http://www.extension .org/pages/30367/abrc2010-interactions-between-nosema-mi crospores-and-a-neonicotinoid-on-honey-bees.

Mullin, C. A., M. Frazier, J. L. Frazier, S. Ashcraft, and R. Simonds. 2010. High Levels of Miticides and Agrochemicals in North American Apiaries: Implications for Honey Bee Health. *PLoS ONE* 5(3): e9754. doi:10.1371/journal .pone.0009754.

Pettis, J. S., E. M. Lichtenberg, M. Andree, J. Stitzinger, R. Rose, and D. vanEngelsdorp. 2013. Crop Pollination Exposes Honey Bees to Pesticides Which Alters Their Susceptibility to the Gut Pathogen *Nosema ceranae*. *PLoS ONE* 8(7): e70182. doi:10.1371/journal.pone.0070182.

Pettis, J. S., D. vanEngelsdorp, J. Johnson, and G. Dively. 2011. Pesticide Exposure in Honey Bees Results in Increased Levels of the Gut Pathogen *Nosema*. *Naturwissenschaften.* doi:10.1007/s00114-011-0881-1.

Steingraber, S. 2011. *Raising Elijah: Protecting Our Children in an Age of Environmental Crisis.* Philadelphia: Da Capo.

US Department of Agriculture. 2012. Report on the National Stakeholders Conference on Bee Health. http://www.usda.gov/documents/ReportHoneyBeeHealth.pdf.

Vidau, C., M. Diogon, J. Aufauvre, R. Fontbonne, and B. Viguès. 2011. Exposure to Sublethal Doses of Fipronil and Thiacloprid Highly Increases Mortality of Honeybees Previously Infected by *Nosema ceranae*. *PLoS ONE* 6(6): e21550. doi:10.1371/journal.pone.0021550.

Winter, C. K., and J. Katz. 2011. Dietary Exposure to Pesticide Residues from Commodities Alleged to Contain the Highest Contamination Levels. *Journal of Toxicology.* Article ID 589674. doi:10.1155/2011/589674.

Wu, J. Y., C. M. Anelli, and W. S. Sheppard. 2011. Sub-Lethal Effects of Pesticide Residues in Brood Comb on Worker Honey Bee *(Apis mellifera)* Development and Longevity. *PLoS ONE* 6(2): e14720. doi:10.1371/journal.pone.0014720.

Zhu, W., T. Reluga, and J. Frazier. 2011. From Subtle to Substantial: A Stage-Structured Matrix Population Model for Predicting Combined Roles of Nutrition and Pesticides on Honey Bee Colony Health. Poster presented at the Entomological So-

ciety of America meeting, Reno, NV, November 14. http://esa
.confex.com/esa/2011/webprogram/Paper57792.html.

Zhu, W., D. R. Schmehl, C. A. Mullin, and J. L. Frazier. 2014.
Four Common Pesticides, Their Mixtures and a Formulation
Solvent in the Hive Environment Have High Oral Toxicity to
Honey Bee Larvae. *PLoS ONE* 9(1): e77547. doi:10.1371
/journal.pone.0077547.

5. VALUING NATURE

Brittain, C., N. Williams, C. Kremen, and A-M Klein. 2013.
Synergistic Effects of Non-*Apis* Bees and Honey Bees for
Pollination Services. *Proceedings of the Royal Society B* 280.
doi:10.1098/rspb.20122767.

Bumblebee Conservation Trust. http://bumblebeeconservation.org.

Chuck, H. 2012. New Almond Promises Independence from
Bees. *Business Journal* (February 9). http://www.thebusiness
journal.com/news/agriculture/770-new-almond-promises-in
dependence-from-bees.

David, E. 2011. Bee-Free Almond Tree. *Growing Produce* (November 11). http://www.growingproduce.com/article/23991
/bee-free-almond-tree.

Environmental Quality Incentive Program, USDA. How NRCS Is
Helping Pollinators. http://www.nrcs.usda.gov/wps/portal/nrcs
/detail/national/plantsanimals/pollinate/help/?cid=nrcsdev11
_001103.

Environmental Working Group. 2012. Farm Subsidy Database.
http://farm.ewg.org/region.php?fips=00000.

Food Alliance. http://foodalliance.org.

Garibaldi, L., I. Steffan-Dewenter, C. Kremen, J. Morales, R. Bommarco, S. Cunningham, L. Carvalheiro, et al. 2011. Stability

of Pollination Services Decreases with Isolation from Natural Areas despite Honey Bee Visits. *Ecology Letters* 14: 1062–1072.

Garibaldi, L. A., I. Steffan-Dewenter, R. Winfree, M. A. Aizen, R. Bommarco, S. A. Cunningham, C. Kremen, et al. 2013. Wild Pollinators Enhance Fruit Set of Crops regardless of Honey Bee Abundance. *Science* 339: 1608–1611.

Integrated Crop Pollination. http://www.icpbees.org.

Karthik, D., B. Kate, J. Waterman, P. Bailis, and M. Welsh. 2011. Programming Micro-Aerial Vehicle Swarms with Karma. *SenSys* 11: 121–134.

Klein, A-M, C. Brittain, S. D. Hendrix, R. Thorp, N. Williams, and N. Kremen. 2012. Wild Pollination Services to California Almond Rely on Semi-Natural Habitat. *Journal of Applied Ecology*. doi:10.1111/j.1365-2664.2012.02144.x.

Klinkenborg, V. 2013. Lost in the Geometry of California's Farms. http://www.nytimes.com/2013/05/05/opinion/sunday/lost-in-the-geometry-of-californias-farms.html.

Local Food Plus. http://www.localfoodplus.ca.

Morandin, L. A., and M. L. Winston. 2005. Wild Bee Abundance and Seed Production in Conventional, Organic, and Genetically Modified Canola. *Ecological Applications* 15: 871–881.

———. 2006. Pollinators Provide Economic Incentive to Preserve Natural Land in Agroecosystems. *Agriculture Ecosystems and Environment* 116: 289–292.

Morandin, L. A., and C. Kremen. 2012. Bee Preference for Native versus Exotic Plants in Restored Agricultural Hedgerows. *Restoration Ecology*. doi:10.1111/j.1526-100x.2012.00876.x.

———. 2013. Hedgerow Restoration Promotes Pollinator Populations and Exports Native Bees to Adjacent Fields. *Ecological Applications* 23: 829–839.

Potts, S. G., J. C. Biesmeijer, and C. Kremen. 2010. Global Pollinator Declines: Trends, Impacts, and Drivers. *Trends in Ecology and Evolution* 25: 45–353.

US Department of Agriculture. 2008. Using Farm Bill Programs for Pollinator Conservation. Natural Resources Conservation Service, Technical Note no. 78. http://plants.usda.gov/pollinators/Using_Farm_Bill_Programs_for_Pollinator_Conservation.pdf.

———. 2010. Self-Pollinating Almonds Key to Bountiful Harvests. *Agricultural Research Magazine* (April). http://www.ars.usda.gov/is/pr/2010/100406.htm.

Velez, L. I., S-Y. Feng, C. S. Goto, and F. L. Benitez. 2011. Dangerous Drug Interactions. *Emergency Medicine Reports* 32(21): 249–261.

6. BEES IN THE CITY

CityFarmBoy.com. http://www.cityfarmboy.com.

Frankie, G. W., R. W. Thorp, M. H. Schindler, B. Ertter, and M. Przybylski. 2002. Bees in Berkeley? *Fremontia* 30: 50–61. http://www.cnps.org/cnps/publications/fremontia/Fremontia_Vol30-No3and4.pdf.

Great Pollinator Project. http://greatpollinatorproject.org.

Great Sunflower Project. http://www.greatsunflower.org.

Growing Power. http://www.growingpower.org.

Hives for Humanity. http://hivesforhumanity.com.

Kearns, C. A., and D. M. Oliveras. 2009. Environmental Factors Affecting Bee Diversity in Urban and Remote Grassland Plots in Boulder, Colorado. *Journal of Insect Conservation* 13: 655–665.

Ladner, P. 2011. *The Urban Food Revolution*. Vancouver: New Society Publishers.

Matteson, K. C., J. S. Ascher, and G. A. Langellotto. 2008. Bee Richness and Abundance in New York City Urban Gardens. *Annals of the Entomological Society of America* 101: 140–150.

Merrick, J. 2012. How Do-Gooders Threaten Humble Bee. *Independent* (October 28). http://www.independent.co.uk/envi ronment/nature/how-dogooders-threaten-humble-bee -8229460.html.

Tommasi, D., A. Miro, H. A. Higo, and M. L. Winston. 2004. Bee Diversity and Abundance in an Urban Setting. *Canadian Entomologist* 136: 851–869.

Two Block Diet. http://twoblockdiet.blogspot.ca.

Winston, M. L. 1997. *Nature Wars: People vs. Pests*. Cambridge, MA: Harvard University Press.

7. THERE'S SOMETHING BIGGER THAN PHIL

American Apitherapy Society. http://www.apitherapy.org.

Brooks, M., and C. Reiner. 1997. *The 2000 Year Old Man in the Year 2000*. New York: Cliff Street Books.

Buxton, S. 2004. *The Shamanic Way of the Bee*. Rochester, VT: Destiny Books.

Castro, H. J., J. I. Mendez-Lnocencio, B. Omidvar, J. Omidvar, J. Santilli, H. S. Nielsen Jr., A. P. Pavot, and J. R. Richert. 2005. A Phase I Study of the Safety of Honeybee Venom Extract as a Possible Treatment for Patients with Progressive Forms of Multiple Sclerosis. *Allergy and Asthma Proceedings* 26: 470–476.

Hood, J. L., A. P. Jallouck, N. Campbell, L. Ratner, and S. A. Wickline. 2013. Cytolytic Nanoparticles Attenuate HIV-1 Infectivity. *Antiviral Therapy* 19: 95–103.

Horn, T. 2005. *Bees in America: How the Honey Bee Shaped a Nation.* Lexington: University Press of Kentucky.

Pius XII. 1948. Speech given on November 27. http://www.catholicculture.org/culture/library/view.cfm?recnum=3813.

Seeds, N. 2013. http://redmoonmusings.com/bee-blessings/sacred-beekeeping.

Wesselius, T., D. J. Heersema, J. P. Mostert, M. Heerings, F. Admiraal-Behloul, A. Talebian, M. A. Van Buchem, and J. De Keyser. 2005. A Randomized Crossover Study of Bee Sting Therapy for Multiple Sclerosis. *Neurology* 65: 1764–1768. doi:10.1212/01.wnl.0000184442.02551.4b.

Wynn, K. 2013. The Land of Bad Milk and Fake Honey. *New Zealand Herald* (August 25). http://www.nzherald.co.nz/nz/news/article.cfm?c_id=1&objectid=11113964.

8. ART AND CULTURE

"Beetalker: The Secret World of Bees." 2006. *The Nature of Things.* CBC television series.

Crane, E. 1983. *The Archaeology of Beekeeping.* London: Duckworth.

Dyck, A. 2011. Interview in the *Mason Journal.* http://www.mason-studio.com/journal/2011/10/interview-with-aganetha-dyck-canadian-visual-artist/.

Ellis, H. 2004. *Sweetness and Light.* New York: Harmony Books.

Harpo, S. "I'm a King Bee." http://www.youtube.com/watch?v=XWLvm11MAaM.

Plath, S. 1965. "Wintering." In *The Collected Poems.* http://www.americanpoems.com/poets/sylviaplath/1461.

Virgil. 29 BCE. *Georgics.* http://www.poetryintranslation.com /PITBR/Latin/VirgilGeorgicsIV.htm.

Waters, M. Honey Bee. http://www.youtube.com/watch?v=jFBq2n W72OI.

Winston, M. L., and G. Lotenberg. 2009. Abandoning Abandon: Dancing Science. *Vancouver Observer* (February 6). http:// www.vancouverobserver.com/blogs/methodscreation/2009/02 /06/abandoning-abandon-dancing-science.

Wyman, M. 2004. *The Defiant Imagination.* Vancouver: Douglas and McIntyre.

Yeats, W. 1888. "The Lake Isle of Innisfree." Poetry Foundation. http://www.poetryfoundation.org/poem/172053.

9. BEING SOCIAL

America*Speaks.* 2006. Unified New Orleans Plan. http://ameri-caspeaks.org/projects/case-studies/unified-new-orleans-plan/.

———. 2010. Our Budget, Our Economy, National Discussion. http://americaspeaks.org/projects/topics/budgeting/america speaks-our-budget-our-economy-2/.

Bloch, G., H. Shpigler, D. E. Wheeler, and G. E. Robinson. 2009. Endocrine Influences on the Organization of Insect Societies. In *Hormones, Brain and Behavior,* 2nd ed., vol. 2, ed. D. W. Pfaff, A. P. Arnold, A. M. Etgen, S. E. Fahrbach, and R. T. Rubin, 1027–1068. San Diego: Academic Press.

Cohen, E. 2013. Understanding and Conquering Technology Overload. Online course at West Virginia University. http:// wvucommmooc.org/c1weekfour/day-three/.

Leighninger, M., and B. Bradley. 2006. *The Next Form of Democracy: How Expert Rule Is Giving Way to Shared Governance— and Why Politics Will Never Be the Same.* Nashville: Vanderbilt University Press.

Lukensmeyer, C. J., and W. Jacobson. 2013. *Bringing Citizen Voices to the Table: A Guide for Public Managers*. San Francisco: Jossey-Bass.

Nabatchi, T., J. Gastil, M. G. Weiksner, and M. Leighninger. 2012. *Democracy in Motion: Evaluating the Practice and Impact of Deliberative Civic Engagement*. New York: Oxford University Press.

Page, R. E. 2013. *The Spirit of the Hive*. Cambridge, MA: Harvard University Press.

Punnett, E. N., and M. L. Winston. 1989. A Comparison of Package and Nucleus Production from Honey Bee (*Apis mellifera* L.) Colonies. *Apidologie* 20: 465–472.

Seeley, T. D. 2010. *Honeybee Democracy*. Princeton, NJ: Princeton University Press.

Simons D., and C. Chabris. 1999. Selective Attention Test. http://www.youtube.com/watch?v=vJG698U2Mvo.

Surowiecki, J. 2005. *The Wisdom of Crowds*. New York: Anchor Books.

Winston, M. L., and L. A. Fergusson. 1985. The Effect of Worker Loss on Temporal Caste Structure in Colonies of the Honey Bee (*Apis mellifera* L.). *Canadian Journal of Zoology* 63: 777–780.

Zinger, D. 2013. 39 Ways to Improve Human Organizations, Work, and Engagement. http://www.davidzinger.com/wp-content/uploads/Waggle-by-David-Zinger1.pdf.

10. CONVERSING

Bayles, D., and T. Orland. 1993. *Art and Fear: Observations on the Perils (and Rewards) of Artmaking*. Santa Cruz, CA: Image Continuum.

Beggs, K. T., K. A. Glendining, N. M. Marechal, V. Vergoz, I. Na-kamura, K. N. Slessor, and A. R. Mercer. 2007. Queen Phero-mone Modulates Brain Dopamine Function in Worker Honey Bees. *Proceedings of the National Academy of Sciences USA* 104: 2460–2464.

Bohm, D. 1996. *On Dialogue*. New York: Routledge.

Clarke, D., H. Whitney, G. Sutton, and D. Robert. 2013. Detection and Learning of Floral Electric Fields by Bumblebees. *Science* 340: 66–69.

Davies, R. 1979. Quoted in a letter to Gordon Roper, March 29, 1979, in *For Your Eyes Only: The Letters of Robertson Davies*, ed. J. S. Grant. New York: Penguin.

Grozinger, C. M., N. M. Sharabash, C. W. Whitfield, and G. E. Robinson. 2003. Pheromone-Mediated Gene Expression in the Honey Bee Brain. *Proceedings of the National Academy of Sciences USA* 100:14519–14525. doi:10.1073/pnas.2335884100.

Ideas. 2012. "Dancing in the Dark: The Intelligence of Bees." Ca-nadian Broadcasting Corporation, Radio One, June 12.

Jowett, B. 1903. *The Four Socratic Dialogues of Plato*. Oxford: Clarendon.

Limb, C. J., and A. R. Braun. 2008. Neural Substrates of Spon-taneous Musical Performance: An fMRI Study of Jazz Im-provisation. *PLoS ONE* 3(2): e1679. doi:10.1371/journal.pone.0001679.

McEwan, I. 1999. *Amsterdam*. Toronto: Random House Canada.

Page, R. E. 2013. *The Spirit of the Hive*. Cambridge, MA: Har-vard University Press.

Seltzer, L. J., A. R. Prososki, T. E. Ziegler, and S. D. Pollak. 2012. Instant Messages vs. Speech: Hormones and Why We Still

Need to Hear Each Other. *Evolution and Human Behavior* 33: 42–45.

Slessor, K. N., M. L. Winston, and Y. LeConte. 2005. Pheromone Communication in the Honeybee (*Apis mellifera* L.). *Journal of Chemical Ecology* 31: 2731–2745.

Winston, M. L., and K. N. Slessor. 1992. The Essence of Royalty: Honey Bee Queen Pheromone. *American Scientist* 80: 374–385.

Zimmer, C. 2012. The Secret Life of Bees. *Smithsonian Magazine*. http://www.smithsonianmag.com/science-nature/The-Secret-Life-of-Bees.html#ixzz2QlTznbNi.

11. LESSONS FROM THE HIVE

Coal County Beeworks. http://www.eri.eku.edu/coal-country-beeworks.

Jutras, L. 2013. Why There's Nothing Wrong with Anthropomorphism. *Toronto Globe and Mail* (August 2). http://www.theglobeandmail.com/life/why-theres-nothing-wrong-with-anthropomorphism/article13584426/.

Mooallem, J. 2013. A Child's Wild Kingdom. *New York Times* (May 4). http://www.nytimes.com/2013/05/05/opinion/sunday/a-childs-wild-kingdom.html?hp&_r=0.

———. 2013. *Wild Ones: A Sometimes Dismaying, Weirdly Reassuring Story about Looking at People Looking at Animals in America.* New York: Penguin.

Wilson, E. O. 1984. *Biophilia.* Cambridge, MA: Harvard University Press.

Acknowledgments

I am grateful to Andrew Nikiforuk, who first encouraged me to write a book about what we can learn from bees. I also am fortunate to live in an urban neighborhood with sixteen coffee shops within two blocks of our apartment, and especially appreciate the coffee and hospitality of my favorite neighborhood locals, where much of this book was written, particularly Milano, Revolver, and Lost + Found.

As always, I am grateful to Harvard University Press and above all my longtime editor and friend Michael Fisher, whose advice and editorial wisdom contributed greatly to the manuscript. Lisa Roberts designed a superb book cover, Carol Hoke provided careful copyedits, and Deborah Grahame-Smith ably shepherded the book through to publication. Three anonymous reviewers provided perceptive and much-appreciated feedback.

Bee Time reflects a lifetime of conversations and experiences with students, colleagues, beekeepers, and interested friends and family who truly are far too numerous to mention but added profoundly to this book over many decades. For their contributions and valued relationships, I am deeply, deeply, deeply grateful.

It is a singular pleasure to acknowledge Lori Bamber. She has been integral at every stage in providing support and encouragement, incisive and stellar editing, and just the right insights to guide me out of the many logjams that plague a writer's journey. Lori is my soul mate in ways beyond what I ever imagined, and her endless capacity for love and her bottomless well of kindness continue to be a daily reminder of the richness of creation.

CHAPTER ACKNOWLEDGMENTS

I am grateful to Bruce Archibald, who provided background references for and advice about the evolutionary history of bees for Chapter 1. In addition, it's a pleasure to acknowledge the many beekeepers who took considerable time to talk with me about honey, judging, and their bees, the subject of Chapter 2. Brian Fredrickson and Virginia Webb shared their passion for artisanal honey, and both were a great joy to chat with. Bob Wellemeyer, I. Bart Smith, and Ann Harman kindly answered my many questions about the Eastern Apicultural Society (EAS) honey show and let me sit in while they judged some very fine honeys. I am also grateful to Brian Marcy for arranging my EAS honey show experiences, and I truly appreciate the warm hospitality extended by the EAS and the interest shown by the hundreds of beekeepers who attended EAS and shared their enthusiasm and their own lessons learned from bees.

I also am grateful for information about the darker side of honey provided by William Hogan of the US Department of Justice and the US Attorney's Office, District of Northern Illinois. In addition, I wish to express my thanks to Heather Higo, who formerly directed my research laboratory. One wonderful afternoon we reminisced about our bee yards, and Heather reminded me of a raft of stories. She helped me to recall the many, many delightful days we spent with students and landowners in the Fraser Valley, telling yarns and learning from each other.

The introductory section of Chapter 2 was originally published in a somewhat different form as "Two Bottles of Mead" in *Bee Culture* magazine. I appreciate the writing platform the magazine provided for me for about ten years in my regular bee culture column, "From Where I Sit." Many of the lessons I've learned from bees began to be articulated through that outlet.

Chapter 3 benefited from the contributions of Howard Husock, high school friend and journalist who visited our research team in French Guiana to research a story at the same time as Ed Zuckerman. Howard refreshed my memory about a number of incidents during that trip and provided the reference to Solman and Friedman's book on the business debacle of killer bee honey.

Chapter 4 profited greatly from a generous interview with Chris Mullins. I also appreciate observations from David Fischer, Maryann Frazier, James Frazier, Yves LeConte, and Jeff Pettis. In addition, Dr. Howard Koseff kindly provided information on drug interactions. Tom Hayden, a fine science journalist and consummate teacher, generously read an early draft and provided much-appreciated encouragement.

For Chapter 5 I am particularly grateful to Claire Kremen and Claire Brittain, each of whom arranged separate productive and enjoyable visits to the Central Valley of California and provided numerous insights into wild bees and almond pollination. Tyler and Drew Scofield were generous with their time and produce from their farm, and my visit with them was invaluable in understanding almond growing. Neal Williams, Amber Sciligo, Leithen M'gonigle, and Franz Niederholzer also shared their extensive wisdom about and experience with wild bees, and I much appreciate their enthusiasm for their research and outreach projects.

I enjoyed a long visit over tea and honey with Allen Garr, reflecting on urban beekeeping, discussed in Chapter 6. Julia and Sarah Common and Yvonne Yanciw were most generous with their time and hospitality in guiding me through Hives for

Humanity, and chef Dana Hauser and Kerrie Bowers similarly toured me through Vancouver's Fairmont Waterfront apiary. James Cane, Paul van Westendorp, and Mylee Nordin assisted with data on both wild and managed bees in cities.

I'm particularly grateful to Nikiah Seeds for sharing her thoughts about shamanic beekeeping and to Peter Molan for answering a number of questions about manuka honey, the subjects of Chapter 7. For Chapter 8 I appreciate conversations about art, dance, and poetry that illuminated aspects of bees as muse, especially with Aganetha Dyck, Gail Lotenberg, and Renee Saklikar.

Talking about bees again with Rob Page, after many years during which our paths had not crossed, was a wonderful reminder of Rob's keen intellect and passion for science. And at the Eastern Apicultural Society meeting in August 2013 I spent a delightful few hours with Tom Seeley as well, enjoying his passion for bees and science. I also appreciate suggestions from Susanna Haas Lyons and Shauna Sylvester about public participatory processes, which enriched Chapter 9.

My friend and colleague Keith Slessor passed away in 2012, and in my mind I still run enthusiastically to his office when I have some new insight about bees to share. I still appreciate—and miss—his inquiring presence. Chapter 10 also benefited from a long conversation about dialogue with Sandy Heierbacher.

In preparation for writing Chapter 11 I conducted many interviews, asking interviewees what they have learned from bees, and was struck by their invariably rich comments. Most of the following were quoted in the text, but I also appreciate the observations of those not quoted directly: Jim Bobb, Claire Brittain, Aganetha Dyck, Nicky Grunfeld, Dave Hackenberg, Tammy Horn, Hope Johnson, Claire Kremen, Gail Lotenberg, Brian Marcy, Suzanne Matlock, Gene Robinson, Amber Sciligo, Drew and Tyler Scofield, Bart Smith, Marla Spivak, Dave Tarpy, Alice Varon, Bob Wellemeyer, and Neal Williams.

Index

Adee, Richard, 29, 30

adulteration of honey, 30–31, 237

agriculture: CCD's impact on, 60–61; change mechanisms for bee conservation in, 227; codependence in, 225–226; competing visions about, 92; diversity vs. monoculture in, 97, 99, 101, 109; economic value of bees, 6, 14; food production, 77–80, 99, 226, 227 (See also *specific crops*); increasing crop yields, 85–86, 94, 96; industrial farming, 92, 97, 99, 108; introduced crops and livestock, 44; poetry about farming, 159; practices impacting bees, 15, 62, 226, 237–238; subsidized crops, 105; urban, 116–118, 229; wild bees economic impact on, 84–86. *See also* ecosystem services; farmers; USDA; *specific farms*

Agriculture Canada, 84

"Alice." *See* Miro, Alice

Allen, Will, 128

Allergy and Asthma Proceedings, 148

almonds, 87–92, 101

altruism, 13–14, 17, 176–177, 230

American Apitherapy Society (AAS), 142, 147

American Beekeeping Federation (ABF), 68–69

America*Speaks,* 191–193. *See also* crowdsourcing

Ames Farm, 23–25

Amsterdam (McEwan), 217–218

anatomy of bees: climate adaptation by, 7; pollen collecting hairs, 5, 13; specialized structures of workers, 13, 21, 155, 178, 211–212; stingless, 8

antibiotics, 27, 28, 62–63, 71

World Beekeeping Congress,
  25–26
World Honey Show, 25
wound and burn dressings,
  145–146, 152
Wyman, Max, 157, 159, 170

X-Files, The (television series),
  164–165

Yanciw, Yvonne, 126,
  127–128
Yeats, William Butler,
  161–163
Young, Michael, 19–20

Zhu, Wanyi, 71–72
Zinger, David, 195–196
Zuckerman, Ed, 49–50